RESHAPING THE COUNTRYSIDE:
PERCEPTIONS AND PROCESSES OF RURAL CHANGE

RESHAPING THE COUNTRYSIDE:
PERCEPTIONS AND PROCESSES OF RURAL CHANGE

Edited by

Nigel Walford
School of Geography
Kingston University
UK

John Everitt
Department of Geography
Brandon University
Canada

and

Darrell Napton
Department of Geography
South Dakota State University
USA

CABI *Publishing*

CABI *Publishing* is a division of **CAB** *International*

CABI Publishing	CABI Publishing
CAB International	10 E. 40th Street
Wallingford	Suite 3203
Oxon OX10 8DE	New York, NY 10016
UK	USA
Tel: +44 (0)1491 832111	Tel: +1 212 481 7018
Fax: +44 (0)1491 833508	Fax: +1 212 686 7993
Email: cabi@cabi.org	Email: cabi-nao@cabi.org

A catalogue record for this book is available from the British Library, London, UK.

Library of Congress Cataloging-in-Publication Data
Reshaping the countryside : perceptions and processes of rural change
    / edited by Nigel Walford, John Everitt and Darrell Napton.
        p.   cm.
    Includes bibliographical references and index.
    ISBN 0-85199-343-5 (alk. paper)
    1. Sociology, Rural. 2. Rural development. 3. Social change.
    I. Walford, Nigel. II. Everitt, John C. III. Napton, Darrell
    Eugene.
    HT421.R38   1999
    307.72--dc21   98-52361
                                                                        CIP

ISBN 0 85199 343 5

Printed and bound at the University Press, Cambridge, from copy supplied
by the editors.

Contents

Contents

Contributors

Bill Ashton,
Fredericton,
New Brunswick,
Canada

Johnathan Bascom
221 Brewster Building,
Department of Geography,
East Carolina University,
Greenville,
North Carolina 27858
USA

Kenneth B. Beesley,
Rural Research Centre,
Nova Scotia Agricultural College,
Truro,
Nova Scotia,
Canada B2N 5E3

John Everitt,
Department of Geography,
Brandon University,
Brandon,
Manitoba,
Canada R7A 6A9

Robert Gant,
School of Geography,
Kingston University,
Penrhyn Road,
Kingston upon Thames,
Surrey KT1 2EE
UK

Alison M. Gill,
Simon Fraser University,
Burnaby,
British Colombia,
Canada V5A 1S6

Ray D. Bollman,
Agriculture Division/Division de
l'Agriculture,
Statistics Canada/Statistique Canada,
Ottawa,
Ontario,
Canada K1A 0T6

Ian Bowler,
Department of Geography,
University of Leicester,
Leicester,
Leicestershire LE1 7RH
UK

Judy Clark,
Department of Geography,
University College London,
26 Bedford Way,
London WC1H 0AP
UK

Philip Lowe,
Centre for Rural Economy,
Department of Agricultural Economics,
University of Newcastle,
Newcastle upon Tyne NE1 7RU
UK

Steve Martin,
Local Government Centre,
Warwick Business School,
University of Warwick,
Coventry CV4 7AL
UK

Richard Gordon,
221 Brewster Building,
Department of Geography,
East Carolina University,
Greenville,
North Carolina 27858
USA

Keith H. Halfacree,
Department of Geography,
University of Wales Swansea,
Singleton Park,
Swansea SA2 8PP
UK

Catherine Lockwood,
Department of Geography,
University of Minnesota,
Minneapolis 55455
USA

Bill Reimer,
Department of Sociology and
Anthropology,
Concordia University,
1455 Boul de Maisonneuve,
Montréal,
Québec
Canada

Susanne Seymour,
Department of Geography,
University of Nottingham,
University Park,
Nottingham,
Nottinghamshire NG7 2RD
UK

Tyrel G. Moore,
Department of Geography and Earth
Sciences,
University of North Carolina at
Charlotte,
Charlotte,
North Carolina 28223
USA

John T. Morgan,
Department of Geography,
Emory and Henry College,
Emory,
Virginia 24327
USA

Darrell Napton,
College of Arts and Science,
Department of Geography,
Box 504,
Scoby Hall,
South Dakota State University,
Brookings,
South Dakota 57007-0648
USA

Charles Watkins,
School of Geography,
University of Nottingham,
Nottingham
Nottinghamshire NG7 2RD
UK

Brian Short,
School of Cultural and Community
Studies,
University of Sussex,
Brighton
East Sussex BN1 9RQ
UK

Nigel Walford
School of Geography
Kingston University,
Penrhyn Road,
Kingston upon Thames
Surrey KT1 2EE
UK

Neil Ward,
Department of Geography,
University of Newcastle,
Newcastle upon Tyne NE1 7RU
UK

Preface

Since the first joint meeting of rural geographers from Canada, the United States and United Kingdom in 1991, strengthening links have developed between the three groups of researchers. These links have proved mutually supportive with collaborative research and exchange visits between institutions as tangible outcomes of the association. In August 1995, the second international conference took place in North Carolina and took as its broad theme rural systems and geographical scale. A companion volume of papers has been published with the title *Agricultural Restructuring and Sustainability: A Geographical Perspective*, which concentrates on one of the specific themes to emerge from the conference.

This volume is based on the second of the main themes, which relates to economic and social changes within the broader rural system. The original papers under this theme have all been revised and updated to reflect intervening research and empirical information. The chapters are organised according to three sub-themes: (i) new countrysides; (ii) socio-economic transformations; and (iii) policy and development responses. This division reflects the manner in which the processes of social and economic change have reshaped the countryside and perceptions of the rural environment have altered. A range of geographical contexts and spatial scales are used in the chapters to illustrate the operation and outcomes of these processes. The reference list integrates literature sources across all the chapters and reflects the contributors' review of research. An index helps to highlight the issues brought out in the individual chapters within the common theme of contemporary countryside change. The book presents a broad comparative examination of the processes and perceptions of this change and offers an analysis and interpretation of these processes.

It is hoped that this book will prove to be useful for undergraduate and postgraduate students in geography and related discipline areas as well as for those whose research interests lie in the area of rural change. The editors are grateful for the help that they have received from the contributors and the organisations representing Rural Geography in Canada, the UK and the US. We would particularly like to thank Claire Ivison for her unstinting patience in redrafting the various maps and diagrams into a common style.

Nigel Walford, John Everitt and Darrell Napton

March 1999

Chapter 1

Continuity and Change in the Developed Countryside

Darrell Napton, Nigel Walford and John Everitt

Introduction

The idyllic appeal of the countryside has near mythical power, drawing to it every year hundreds of thousands of families and individuals in search of a higher quality of life. Further millions visit the countryside on weekend outings or vacations. Yet for many people the countryside is relatively unknown. Until recently people knew, or thought they knew, what it comprised both in terms of how it appeared and of the activities it could sustain. During the past few decades, however, what some people regarded as its immutable characteristics have changed, if not out of all recognition, at least substantially, from traditionally accepted preconceptions. The countryside, as with other landscapes, exists and is created within particular cultural and physical contexts. It is a mosaic made up of many geographies, the result of the interplay of local resources and people, national policies and large-scale economic processes. The mosaic becomes a kaleidoscope when the processes of change accelerate or produce abrupt outcomes. There is not one countryside, but many. This applies with respect both to those people who live or spend at least part of their lives in the countryside and to those who place academic interpretations on the activities undertaken, the environmental characteristics and people found in such areas.

The collection of papers in this edited volume, having been written by researchers from Canada, the United Kingdom and the United States, provide comparative and contrasting impressions of how the countryside has changed in these three countries. In some instances, these changes are relatively recent, whereas in others they have occurred over a longer time period. The spatial scales range from local areas to entire nations, and the methodologies represented range from empirical to theoretical, and from traditional to radical. This volume extends the continuing debate about the nature and outcome of change in the countryside in developed economies. The authors explore issues of continuity and change associated with the operation of demographic, socio-economic, and political processes as they act upon and change the countryside. The mosaic of the countryside aptly depicts this mosaic of ideas, theories and methodologies.

The Geographical Mosaic of the Countryside

While the contributions are divided into three sections, they nevertheless overlap and complement each other. The varied parts of the mosaic produce a single picture. This picture, the predominant theme of the following chapters, is one of change. In recent years, certain rural areas have begun to change after decades of general stability. In some areas, the changes were gradual, while in others they were as rapid as the wholesale deindustrialisation of a region in response to restructuring of the mining industry. Some changes are the result of incremental national policy adjustments and others happen abruptly, as in a grassroots rejection of the prevailing system of property ownership and access to land. Beginning in the 1970s, many rural areas, having faced population decline for several decades, started to grow, largely because access to urban areas or the presence of local amenities made them attractive to retirees, footloose workers and tourists. Other rural areas, however, continued grappling with economic stagnation and decline. The mosaic that is the countryside is taking different shapes. These chapters represent the varied ways we seek to understand the new rural geography that is emerging.

Processes of Countryside Change

Three major processes of countryside change are addressed in this volume: devolution of political and economic responsibility, industrial restructuring and changes in mobility and spending. In reality, these processes often work in concert with each other and with a variety of other elements and, of course, take different paths within each local area. Understanding the countryside may be analogous to understanding a symphony. There are major themes, but each is played out with a range of local variations. One of the strengths of this volume is its examination of a mixture of thematic, or general theoretical processes, and its exploration of how these play themselves out in a number of specific areas.

Political and economic responsibility has been devolving from national to lower levels of governance. Long-standing government programmes have been recently changed or eliminated. Farmers have responded to reduced and changed government oversight by diversifying (Chs. 2 and 3), and by expanding and focusing their efforts on their best soils (Ch. 5). A new environmental protection body for rivers in the UK has attempted to involve a wider range of groups in the development and monitoring of policy for managing water systems and dealing with conflicts which may arise (Ch. 4). To become eligible for provincial assistance, towns are being asked to allow community members to help plan (Ch.

13) and community based economic development is being encouraged (Chs. 14 and 15).

Industrial restructuring is a second process causing countryside change. Some tobacco farmers have created 'super farms' that control a significant percentage of the market and have labour demands that can only be met by Mexican migrants (Ch. 5). The technological and corporate changes associated with agricultural restructuring generally require fewer farmers or farm workers and thus leave a surplus labour force in the countryside (Ch. 7). Industries seeking low-wage workers have often moved to areas with a large surplus of labour. The resulting jobs allow a low-density, rural commuting work force to trade higher wages for a countryside lifestyle. Sometimes restructuring forces a community to make a transition to a new economic base such as from forestry to tourism (Ch. 13). For areas that are isolated, specialised or devoid of other opportunities, industrial restructuring can be particularly traumatic as Moore found when the mines in an entire West Virginia coal mining region closed and there were no alternative jobs. Sometimes the results of restructuring are not readily apprehended (Ch. 12): when the British brewing industry restructured, pubs were sold to pub companies that required profit maximisation, which in turn required pub managers to cater more to visitors than to villagers.

The third major process of countryside change is the movement of people and their money. Normally, young people and young families are more prone to move. Their departure from some communities and arrival in others may alter the demographic structure and as a result, the demand placed upon community resources. The communities to which they move become younger and more family oriented (Ch. 8) and the communities from which they depart become older (Ch. 11). Rural and urban people, who staked their future in the rural-urban fringe, have created an urbanised version of the countryside that provides a lifestyle that most of them enjoy (Ch. 8). Beyond the urban fringe, towns that previously had an economic base not directly dependent upon the city have been recreated as dormitory communities (Ch. 13). Beyond these are large expanses of rural countryside. Other than their relative inaccessibility and general dependence upon local resources, these areas have little in common.

The impact of new residents may result in more or less business and changed functions for some businesses (Ch. 12). In some areas, new residents are considered newcomers for their entire lives and are never truly accepted into the community (Ch. 16). Newcomers may also find themselves excluded from community decisions by the traditional elite (Ch. 13). Often the very presence of newcomers with their different lifestyle changes the community in subtle but significant ways. For example, Bowler and Everitt found that, because newcomers typically drink less than native villagers, pub managers must begin catering for visitors and putting greater emphasis on food to stay in business. Meanwhile, village natives report that the pub is no longer the hub of local social interaction that it had been traditionally.

Consequences of Change

Restructuring of the primary economic sectors, changing national and local government policies, and migration and tourism have resulted in an altered countryside. The consequences of this range from the passing of old ways of life to new geographies of agriculture, industry, lifestyle, wealth and poverty. These in turn may result in new productive lifestyles or frustration and possibly political action. Political changes and restructuring in agriculture are beginning to alter agricultural geography at all geographical scales. Some farmers are responding by growing new crops or diversifying into non-agricultural enterprises (Chs. 3 and 16). At the regional scale farmers may be specialising in selected crops on larger farms (Chs. 3 and 7) and locally, large farmers may be dominating the market and focusing their investments in specialised knowledge and specialised technology and on the best soils (Ch. 5). Restructuring of the mining sector (Ch. 10) and forestry sector (Chs. 13 and 16) has caused even more dramatic changes in the affected communities. Meanwhile, the rural-urban fringe has become a permanent transition zone that provides lifestyle options that are neither urban nor rural and are highly sought by some (Ch. 8).

New Patterns of Poverty and Wealth

The changed countryside has become a dream come true for some, but for those who are immobile or locked in poverty, it has become a nightmare with deprivation, which may result in tension, stress, frustration and anger. When unemployment levels escalate after the closing of mines (Ch. 10), or the traditional elite deny political access to newcomers (Ch. 13), or poverty and reduced mobility reduce the quality of life of the elderly (Ch. 11), the countryside may become a place of anger and despair. If frustration reaches a critical point, groups may resort to civil disobedience (Ch. 6).

A new mosaic of poverty and wealth is emerging in the countryside. Regions of poverty may emerge when an industry that an entire region depended upon depletes local resources and leaves (Ch. 16) or restructures so that facilities can be closed or operated by fewer workers (Ch. 10). Moore discusses new poverty that resulted from industrial restructuring of coal mining. Resource poor regions or technological change may cause out-migration and leave behind an increasingly impoverished, ageing population (Ch. 11). The immobile suffer the most. This would commonly include the elderly, unskilled workers and single mothers (Chs. 9 and 11). The types and concentrations of rural poverty may vary from place to place (Ch. 9). In Canada, the elderly, workers in primary industries, single mothers and construction workers in inaccessible areas compose the largest groups of rural poor. Each of these populations is concentrated in different areas of rural Canada.

Many parts of the countryside, however, are growing in wealth. Areas with significant numbers of workers displaced by restructuring may attract new industry and prosper (Ch. 7). Skilled, knowledge-based farmers can prosper by either growing larger and specialising, by diversifying, or by focusing on new, high-risk, high-profit ventures (Chs. 3, 5 and 16). Areas that are close to urban centres can become part of the rural-urban fringe, dormitory communities, or offer a pleasing atmosphere for country outings (Chs. 8, 12 and 13) and scenic areas may offer new services employment for skiing, hunting, fishing, sightseeing and other forms of outdoor recreation and tourism (Ch. 16).

What should be, or can be, done to promote economic growth in the poorer parts of the countryside? First, at the national scale, we need to learn who is poor, for what reasons and where they are located (Ch. 9). Second, we need to determine the most effective strategies for eliminating or alleviating poverty that are politically possible. The most effective national investments have been shown to be in human capital (Ch. 4). Investing in prenatal care, and the nutrition and nurture of children from the time of birth to the age of three is particularly critical (Ch. 17). If communities are to take advantage of programmes or matching moneys related to economic development, they must have or educate citizens who have organisation and leadership skills (Ch. 15). Properly targeted programmes can succeed. A small public investment in a Minnesota boreal forest community in the 1970s has become a fabric business with national distribution and 300 employees (Ch. 16). Individuals must also be clever and aggressive in making opportunities, using local resources creatively and realising that living in the countryside may have economic trade-offs (Ch. 16).

Some communities no longer have the will or the resources to survive (Ch. 15). Eventually their decline will make them indistinguishable from the surrounding countryside and their existence, as a separate, viable place will cease. Other impoverished places may have so many residents that compassion or political necessity forces some action. However, it is unclear what action can be taken to eliminate or even significantly alleviate much rural poverty. Parts of the Appalachian Mountains have suffered the effects of a lagging economy for over a century. The private relief efforts started in the 19th century and the public initiatives and programmes of the 20th century were designed to make the region prosperous and more accessible, but these programmes have had limited impact (Ch. 10). Increased accessibility has made it easier for the talented and ambitious to leave and for religious and social organisations to provide assistance.

Countering rural poverty is not straightforward. Caution must be used, moreover, when trying to interpret the countryside, because things are not always as they seem. Minnesota's North Woods appears to be a region of poverty, because most of the people do not have jobs; however, these people have developed a host of adaptive strategies that allow them to maintain a comfortable life (Ch. 16). Changing technology may provide tools to help some communities. Perhaps aspects of the computer and telecommunications industries can be used to increase accessibility to better health care for poor or elderly people (Ch. 11).

New Countrysides

The common preconception of the countryside is that its physical, economic and cultural basis lies with farming and the agricultural industry. Chapters in the first section, New Countrysides, consider how not only the physical but also the social and cultural landscapes of agriculture have been altered. For example, agriculture has been transformed from productivist to post-productivist modes of operation, which have acted as a conduit of cultural and environmental change. Moreover, this section presents evidence that old methods of gathering information may ignore new activities such as environmental stewardship and agricultural diversification.

The origins of the productivist era are sometimes associated with the changes brought about as a result of the wartime emergency measures, which pulled agriculture out of a long period of depression and stagnation. Brian Short and Charles Watkins examine one important source of information about conditions on Britain's farms during the early years of World War II. The National Farm Survey, 1941-43, has recently been released to researchers in the United Kingdom through the Public Records Office, Kew and is potentially a valuable source for establishing the nature of farming at that time. Short and Watkins maintain that the Survey is important not simply as a source of data, but as a forerunner to the ethos of government intervention, which came to be accepted as 'normal' in the post-war years. Such developments generated a 'new countryside' not only with respect to agriculture, but also with regard to the management of the countryside for a range of purposes.

In Chapter 3, Nigel Walford examines the patterns of agricultural land use that resulted from the post-World War II productivist phase of policy and practice in Britain and how it may be changing in the post-productivist era. National agricultural policies designed to increase production and lower food costs resulted in concentration, intensification and specialisation of production. These processes operated through investments in larger machinery, development of scientific knowledge and farm enlargement. Parallel developments also occurred in the United States and Canada. During the 1980s, policies were developed to initiate a restructuring of the farming industry. Walford found that current data sources are not effective at capturing farmers' attempts to respond to the new policies. This is also the case in Canada and the United States. He further suggests that the conceptual definition of a farm may need to be revised to capture the multiple functions which increasing numbers of farms are trying to perform. He concludes from the evidence provided by agricultural land use statistics that farmers' responses to the new policies seem to be changing the agricultural geography of a belt of farms located between the cereal regions of the South-East and the livestock regions of the North-West.

British farmers enjoyed a comparatively privileged position during much of this post-war era. Agriculture was relatively untouched by planning legislation and financial support was provided in order to modernise farm resources and to improve land. Susanne Seymour and her fellow researchers focus on one aspect of environmental monitoring and regulation, the protection of water courses from farm pollution, to consider whether a new moral order has emerged with respect to the formulation of environmental policy. The management of river systems in the UK has experienced several changes in the last 30 years, however those associated with the creation of the National Rivers Authority in 1989 were compounded by the privatization of water supply. The NRA announced that it would adopt a stronger stance in the protection of the UK's river systems against polluters, including the farming community. The authors conclude that the reformist and educational rather than revolutionary stance adopted by the NRA contributed to its lack of success in instilling a new moral order into environmental policy.

US tobacco policies present a striking example in the interaction between farmers, government and the environment. John Morgan presents a case study of how farmers in the Great Valley of South-West Virginia have responded to successive tobacco policies designed to prevent surpluses, maintain high prices and allow small farmers to be competitive with large farmers. Since the 1930s, farmers and the government have established a dynamic equilibrium with farmers finding loopholes in tobacco legislation and developing technologies to increase yields and the government responding with revised laws designed to limit production and support small producers. Today, small tobacco farmers face extinction as the US Congress debates eliminating the tobacco programme because of the health risks associated with tobacco consumption. Larger farms will be able to continue profitable production, but small farmers will not, and there are no viable alternative crops.

Socio-economic Transformations

The second section, Socio-economic Transformations, focuses upon specific social and economic changes confronting rural communities. The types of changes are as diverse as economic growth and economic collapse, struggles to gain access to land and housing, and struggles of the elderly to maintain access and quality of life as ambulatory decline becomes significant. This section comprises a group of papers that concentrate explicitly on issues of welfare, social space and conflict within the countryside. Several of the contributions connect with the contemporary debate about rural poverty and the conditions of disadvantaged groups in the countryside.

Historically, farming was viewed as benign to the countryside and as the proper rural land use. Beginning in the 1970s, the quality of farmers' stewardship began to be questioned. Keith Halfacree reviews Lefebvre's ideas and applies them to British groups that are questioning the access granted to the landless in the countryside. Some of these groups are questioning the validity of private land ownership. Halfacree postulates three models, or alternative futures, for the British countryside. The version promoted by those who currently feel excluded is decentralised and more locally oriented than the competing productivist farming landscape or the rural idyll landscape of small farms and villages.

Before 1970, rural counties in the United States lost people to metropolitan ones, but beginning in the 1970s, rural counties, even ones distant from cities, began to grow. Johnathan Bascom and Richard Gordon investigated the reasons for this rural population 'renaissance' in the largest area of rural counties East of the Appalachian Mountains, the eastern coastal region of North Carolina. Agricultural mechanisation and the transition to corporate and industrial farming dramatically reduced the number of farms and farm workers after 1950. During the 1960s and 1970s, there was a wave of rural industrialisation with businesses searching for areas of surplus workers who were not unionised and were willing to trade higher urban wages for a rural lifestyle. Contributing to the non-farm population growth in coastal North Carolina was in-migration associated with tourism and retirement. Within a fifty-year span, this region made the transition from that of an agricultural economy to a service and manufacturing economy.

How do rural-urban communities differ in terms of their residents' satisfaction with life? Ken Beesley provides answers by reviewing 15 years of research in Southern Ontario. There were no dramatic differences from one urban area to another nor for communities at the edge of large cities compared with small cities. Fringe residents were willing to accept longer commuting journeys and reduced access to goods and services to obtain more housing and space for their families. Some fringe residents were dissatisfied with the level of local services. Satisfaction with life tended to be more related to personal circumstances such as standard of living and family life than with residential choice. Community satisfaction was related to the community as a place to raise children and engage in social activities.

Bill Reimer investigated the types of rural poverty and their spatial concentration in Canada. The elderly constitute the largest group. They are usually unemployed, dependent upon government assistance and are located throughout Canada with significant concentrations in Québec and the Atlantic Provinces. The second largest group of poor people has a head of household employed in a primary occupation, usually farming, and are therefore focused on the farming regions. The third largest group comprises families headed by single mothers. They are most commonly found in the northern parts of most provinces with a secondary concentration in Québec. Poor construction workers, the fourth group, are also most likely to live in remote areas but in different places to the single mothers. A major conclusion is that the poor differ from place to place. Consequently, programmes to help them must be flexible, multi-faceted, spatially targeted and directed to different groups.

The coal fields of southern West Virginia provide an example of what occurs when a rural resource-based economy undergoes deindustrialisation and restructuring. Tyrel Moore shows that while service sector employment absorbed industrial losses in many areas, this did not occur in southern West Virginia. The failure to make that transition was linked to the region's historical total dependence on the coal industry. During the 1980s, mining job losses displaced up to 75% of the work force in some areas. Today, the region is still economically distressed. Moore profiles three developments that symbolise the fragile nature of this poverty stricken region. First, national waste management companies view the region with its combination of abandoned coal mines and desperately poor people as excellent for developing landfill sites. Accepting out-of-state waste would provide jobs and local government revenue, but it would also threaten water supplies. Second, churches from outside the region support ministries as well as provide blankets, clothing, toys and housing rehabilitation and construction. Third, the region has recently been placed on the list of places to visit in a San Francisco travel agency's 'Third World in America' tour.

If we live long enough, we shall all have locomotion difficulties. Robert Gant focuses upon mobility-impaired people in the rural area of the North Cotswolds, England. This area has had out-migration of young people and in-migration of retirees, leaving it as a region of small settlements with reduced community services and a high proportion of elderly. Attempts to use telecommunications as a tool to help the elderly in the Cotswolds can be used as a model for other ageing rural communities in developed nations. Gant is optimistic that computers, the Internet, electronic mail networks and interactive television can increase the quality of life for many housebound elderly and provide them with better and quicker health care options. One significant barrier is that many elderly people do not have the financial resources to purchase these tools. He then presents a model for developing community and travelling telecommunications centres for the elderly poor.

The rural pub has long been a fixture of the British landscape. It provided a place for local social gatherings and an atmosphere where villagers could relax. Ian Bowler and John Everitt discovered that many rural pubs have either dramatically changed or closed in response to four large-scale processes. The brewing industry has consolidated and sold many pubs to pub companies that have remodelled them into drinking houses that cater to families and emphasise meals as much as drink. This capital influx simultaneously reinvigorated many pubs and forced managers to focus on maximum profits. Unprofitable pubs were closed. Secondly, stricter laws governing drinking and driving and food safety resulted in fewer driving customers and higher operating costs. Third, many villages now have high proportions of new residents who moved to the countryside to increase their quality of life. Typically, these newcomers do not patronise pubs as frequently as established villagers, thus forcing the manager to become dependent upon visitors and their preferences. Last, the role of the pub has changed from being a hub of village social interaction into a centre for recreation and entertainment by outsiders. These changes have caused the role of the British pub to change from an object of folk culture to one of popular culture.

Policy and Development Responses

Policy and Development Responses, the final section, includes papers that focus on developmental and socio-economic policy in order to compare the effectiveness of policies with different origins. National agricultural policies developed during the Great Depression and World War II largely ossified the geography of agriculture. Following World War II, agricultural production rapidly increased, but because of changes in technologies, fewer farmers and farm workers were needed. Regions that were dependent upon other natural resources, such as forests and minerals, faced similar situations. As unemployed and underemployed workers migrated and others were stranded, rural communities found or created new economic bases or began to decline. The economic and political processes that were causing these changes normally originated at a larger, more encompassing scale. As a result, national efforts were initiated to assist inappropriately skilled workers and declining communities. Contributors to this section grapple with ways in which policy may increase the quality of life in the countryside at all geographical scales. They provide some initial answers and help clarify the issues by articulating better, more insightful research questions. One recurring emphasis in the chapters is on the contribution of community-based approaches to policy formulation and implementation.

Squamish, British Columbia, is a traditional forest resources town that is now within commuting distance of Vancouver and is on a frequently travelled tourism road. Squamish was forced to develop a community-based tourism plan if it wanted provincial approval for a nearby ski development. Allison Gill recounts the tensions that arose during the planning phase. One of the primary tensions was between newcomers and natives. Eighty per cent of the planning committee members were new residents who had different values from many of the natives. The forest industry, without whose co-operation a plan would fail, declined to participate. This left the traditional elite in a position to argue that the final plan did not represent the entire community. After the plan was submitted, it was ignored by the Metropolitan Council.

Community involvement in rural development also provides the focus for Steve Martin's chapter. He assesses the effectiveness of initiatives promoting community involvement in rural policies, which have been funded at the national and European levels. He concentrates on two of the most important British and European rural development initiatives of recent years, known as Rural Action for the Environment and LEADER1. The broad principles embodied by Rural Action and LEADER1 continue to be widely supported by policy-makers and the future funding of such schemes seems more assured than other areas of public support. However, they continue to pose ideological and practical dilemmas, for example in respect of whether they represent a derogation of responsibility for welfare by state bodies.

Community-based approaches to economic development are emerging as a focus of public policy as federal responsibilities devolve to local levels and national moneys increasingly fund or match local initiatives. In his chapter on community readiness, Bill Ashton evaluates the ingredients that allow some communities to succeed. He concludes that the best way to improve the odds of community success is to focus on the 'getting started' stage. To that end, he develops a self-evaluation framework to help communities determine whether they have the resources to begin Community Economic Development. Ashton also discusses what to do with communities that have neither the will nor the resources to support economic development.

Catherine Lockwood investigated Minnesota's boreal forest, an area of logging in the 19th century that 20th century settlers tried to farm. The soils were too poor and the growing season too short for most profitable agriculture, yet many settlers stayed as the area became reforested. Since the 1960s, there have also been new waves of migrants. In this marginal region, only 60% of the work force have jobs and of these only one-third work full-time. Lockwood investigates who these people are and how they are able to survive. She learned that natives, returning natives and newcomers often have different views of the area and look to different economic strategies to survive. Each group, however, uses adaptive economic strategies that cleverly utilise local resources. Many families depend upon a balance of part-time jobs, bartering and seasonal resource exploitation to create the quality of life that they desire.

Rural Canada has a surplus of persons looking for jobs. The results of this situation include out-migration, higher unemployment and lower incomes than in urban Canada. Ray Bollman investigates how local economic development programmes should focus resources to increase the number of jobs and level of income. His national study of Canada by census region indicates that developing local human capital is a necessary, but not sufficient, condition to support local economic development. The investment with the highest return is prenatal care and providing nutrition and nurturing for children up to the age of three.

Conclusion

The countryside is a multi-faceted place. It looks different to newcomers than it does to natives; it looks different to those who choose to live there than to those who cannot leave. For some it provides opportunities and a pleasing lifestyle where nature is predominant; to others it is a trap that apparently has no release.

The allure of the countryside will continue. Open spaces are larger and more inviting there, and people are fewer and may be perceived as being more hardworking, honest or genuine. For many, the countryside ignites dreams of being close to nature, self-sufficiency, new opportunities and of genuine, earthy communities. Enough of those perceptions are accurate that new dreams are kindled each generation. Each generation, however, is faced with a new countryside. The earth abides, but resources, property, towns, communities and all of the other tiles that compose the mosaic of rural areas are reshaped into a new geography of the countryside.

Chapter 2

Prelude to Modernity: an Evaluation of the National Farm Survey of England and Wales, 1941-43

Brian Short and Charles Watkins

Introduction

In the first half of the twentieth century English agriculture was propelled from an old to a new order by the growth of large-scale administrative and bureaucratic systems of regulation. The requirement for surveillance over the nation's agriculture, which had been felt for generations but was now insistent in the face of the Great War and the following depression, became a central feature. Surveillance yielded information, an essential modern prerequisite for power and control. And as political ideologies changed, so ideas about increasing intervention by the state blossomed within rural England as elsewhere.

Rural survey work had become widely recognised by the late 1930s through the work of the regional survey movement (Matless, 1998); the Land Utilisation Survey (Stamp, 1947; Rycroft and Cosgrove, 1995) and the Farm Management Survey (Murdoch and Ward, 1997). Proposals for a national agricultural survey were put forward by officials in the Ministry of Agriculture (MAF) and agricultural economists in 1938. Discussions were held on the "desirability of a survey which would show which land ought to be reserved for agriculture in connection with any State planning of the utilisation of land for industry, building development, aerodromes etc." Such proposals were not welcomed by some officials, one suggested in March 1939 that "a comprehensive survey would only be needed and could only be justified if the Government were to contemplate a measure of Government control on Socialist or Germanic lines over the utilisation of agricultural land and the operation of individual farmers".[1]

With the onset of war in September 1939 such views were pushed aside and under the Defence (General) Regulation 49, the Minister of Agriculture was empowered to establish County War Agricultural Executive Committees

(CWAECs or 'war ags') in order to increase food production (Murray, 1955). A massive programme to convert pasture and meadowland to arable was instigated by the CWAECs and to assist this plough-up campaign a farm survey was carried out (Stapledon, 1939). Two special surveys of agriculture were carried out in these early years of the war. The first survey, which began in June 1940, was important in its own right, but was in reality little more than a precursor to the second much greater investigation, the National Farm Survey (NFS).

In the first survey farms were classified in terms of their productive state, and between June 1940 and early 1941 about 85% of the agricultural land area of England and Wales was surveyed. The principal concern of this first survey was to encourage the increase of food production. It included several questions, which were not in the later NFS such as requests for information on methods of drainage, fertiliser requirements and vacancies for the Women's Land Army. Summary county reports of this survey survive but most of the original farm records have been lost. The quality of the county reports varied greatly as did the range of questions asked. It was partly the inconsistency of this survey which led to the development of a new national survey.

The idea for a complete NFS, which was described at the time as a 'Second Domesday Book', grew out of this early survey. As well as assisting the CWAECs with their wartime administration, the NFS was intended to form "a permanent and comprehensive record of the conditions on the farms of England and Wales at the time of recording - the compilation of a modern Domesday Book - the ultimate intention being to place the records and maps in the PRO", and to assist in post war planning, advisory and educational work.[2] Initially it was thought that the "uniform form of record" that was required would be put together from information already obtained by Committees.[3]

Already, by September 1940, during the first heavy air raids on London, Ministry of Agriculture officials were setting out proposals for a new survey and Circular 545 of 26 April 1941 initiated the NFS.[4] There is no real evidence that an official end was ever brought to the NFS, but by July 1943 thoughts were turning to the analysis of the collected data by the Advisory Economists attached to the project, although the final summary report did not appear until August 1946 (Ministry of Agriculture and Fisheries, 1946)

The NFS was designed by a Farm Survey Committee which was composed almost entirely of ministry officials and representatives from the CWAECs, although the geographer Dudley Stamp was a key member. Stamp's wide experience of land use studies and his organisation of the Land Utilisation Survey, whose maps and reports were being published at this time, made him an obvious choice as a committee member. His influence can be seen in the emphasis placed on the importance of statistical and cartographic analysis of the NFS, and he certainly considered the maps to be "of fundamental and outstanding importance for any work on rural planning and agricultural reconstruction".[5]

The whole NFS was devised from the outset as a decentralised operation, to be conducted under the auspices of the County War Agricultural Executive Committees (CWAECs) in each locality. Following initial preparations, visits to each farm were undertaken by CWAEC District Committee members, the 'maids-of-all-work' at this time (Rutherford and Bateson, 1946, p. 130). The outcome of such visits was the completion of the Primary (Farm) Return which provided detailed information on tenure, the conditions on the farm, water and electricity supplies, and farm management. Much information was already gathered by the CWAECs from earlier surveys and returns. Paid workers and volunteers were used; most surveyors were local farmers, technical officers and land agents. The information collected was sent to one of eleven Provincial Agricultural Advisory Centres based in university agriculture departments or agricultural colleges. Here the records were checked and matched up with the 4 June 1941 Agricultural Census. This enormous amount of information covered most holdings of 2 ha (5 acres) or more. Woodland was excluded from the National Farm Survey, but the Forestry Commission Census of 1947/9 provides complementary statistical information and maps for all woods of over 2 ha (5 acres) (Watkins, 1984).

The NFS was one of the fundamental components of the intervention by the state in the wartime economy and provides an excellent example of the surveillance held to be necessary in order to improve the productivity of agriculture at a time of potential food scarcity and crisis (Short and Watkins, 1994; Murdoch and Ward, 1997). As Lord Woolton (Minister of Food 1940-1943) later stated "Those who found themselves in 1939 responsible for the conduct of the country in the rapidly changing conditions of war - when facts were more important than precedents - had very little to help them" (Moss, 1991, p. 3). The planning of the survey indeed began at a time of very real crisis, as the Battle of Britain was being fought out in the skies above the fields and farms. It was launched in the weeks following the German invasion of Russia in the summer of 1941 and was being fully implemented by the time of the Japanese attack upon Pearl Harbour in December 1941 which brought America into the war. The survey ended at a time when the tide of war was turning fundamentally in favour of the Allied forces and when, as Churchill later calculated, three-quarters of the world was ranged alongside Britain (Churchill, 1949, p. 510).

In 1994 a project funded by the Economic and Social Research Council (ESRC) was established to investigate the value of this source for the study of agricultural and social conditions in mid-twentieth century England and Wales. The bulk of the NFS records themselves are held at the Public Record Office (PRO), Kew. In this chapter we outline the significance of the NFS and provide an introduction to the quality and research potential of the data and maps which constitute the NFS archive (Short *et al.*, forthcoming 1999).

The National Farm Survey Records

The records of the 1941-43 Survey were recalled from the Provincial Advisory Centres, where they had resided since the war, and assembled by Ministry archivists at Hayes in the late 1950s as five related, but separately packaged, groups of records. The forms were then sent to the Public Record Office at Chancery Lane in 1959, and the maps in 1970, and transferred to the new Kew office in 1976-77. At the PRO there are now the four different sets of forms and a set of maps - and thus theoretically five different elements yielding data for each holding in England and Wales of 2 ha (5 acres) or more. Because the Ministry had been careful to guard the confidentiality of the material - especially the grades relating to the management of the farms - the records were not open to public inspection for a 50-year period, deemed to end in January 1992.

The National Farm Survey archive consists of the following components, numbers 1 to 4 of which constitute PRO class MAF 32. (The references in square brackets are to MAF's own internal reference class for each form.)

1. The Primary Return [B496/EI]. This was based on information collected by the CWAECs supplemented by additional field visits to farms. The Record provides detailed information for each farm under several main headings:

a. Tenure: address of landowner 1941-43; occupancy and multiple holdings; part-time/full-time occupancy and grazing rights.

b. Condition of the farm: generalised soil conditions; farm layout; condition of roads and fences; tied cottages and cottages not on the farm; degree of infestation with weeds or pests, and number and area of derelict fields.

c. Water and electricity: source of water supply to different farm buildings and seasonal shortages, and extent of electrification.

d. Management: a controversial qualitative assessment as to the ability of the farmer, given in terms of an A, B or C grading, with occasionally more detailed plus or minus gradings employed, and reasons given for B or C grading, such as old age, lack of capital or 'personal failings', with space for more detail if the latter, allowing amplification on such issues as physical incapacity, lack of ambition, stupidity, laziness or 'weaknesses of the flesh' such as drunkenness.

2. The Census Return [C47/SSY, p. 1]. The 4 June Return 1941 for each farm - the only year for which such information has ever been released on a farm-by-farm basis rather than as a parish summary. Complete land use, livestock and labour input details are thus discernible. These were completed, as usual, by the farmers themselves, as also were 3 and 4 below.

3. The Horticultural Return [C51/SSY, p. 2]. The 4 June 1941 Census Return (Small Fruit and Horticultural Produce). This gave more detail for each farm on small fruit types, vegetables, flowers, crops being grown under glass, and also stocks of hay and straw.

4. The Supplementary Form [398/SS]. The 4 June 1941 Census Return ('Supplementary Form' concerning Motive Power, Rent and Length of Occupation). This gave more details on labour (regular and part-time, casual, etc.); on the motive power on the holding (tractors and stationary engines, with engine manufacturers' type and hp); rent (amount payable and whether the holding was part-owned and part-rented); and the length of occupation (in years), including parts occupied for different lengths of time.

5. Maps were used to plot the boundaries and fields of each holding. These are now available as PRO MAF 73 and constitute either the Ordnance Survey 1:2,500 sheets reduced to half-size at approximately 1:5,000, or the 1:10,560 sheets. This was exacting work, inevitably done to different standards, with the best having a colour wash over the whole farm area or colouring of farm boundaries, with the farm code reference in black ink. If 1:10,560 sheets were used, then the plot numbers from the relevant 1:2,500 sheets were added to each field. Sometimes areas were transcribed as well. Since this threatened to clutter the map unduly, many sheets had marginalia added with information on ownership and keys to the colour scheme used.

The Survey Mechanism

Our research has shown that the detailed method of survey was far from simple. Figure 2.1 is a representation of the mechanism of the NFS. At the bottom of the diagram are the farmers of England and Wales: the source of the survey information. There are four organisational centres:

a. The Survey was planned and implemented at MAF headquarters in London.
b. The inspection of farms was carried out by staff based at the 62 CWAECs of England and Wales.
c. The Advisory Economists were based at 11 centres in England and Wales and here records were checked and collated.
d. The 4 June Census returns were sent out from the MAF Statistical Branch based at hotels in St Anne's Lancashire. Here farm survey information was checked and matched against the census data.

The movement of forms and parts of forms for checking and copying between the MAF Statistical Branch, the Advisory Centres, the CWAECs and the farmers was exceptionally complicated and resulted in several problems. Our approach was to repopulate the Survey and reconsider the importance of the different elements. One element that can now be identified as being of immense importance is the network of Provincial Agricultural Advisory Centres. These centres maintained professional staffs undertaking research into technical and economic aspects of farming and were crucial to the implementation of the Survey. The Advisory Economists, who had been administering the Farm Management Survey of 2,000 farms since 1936, represented the 'think tanks' of British agriculture.

Correspondence between the Ministry and the Advisory Centres survives to show some of the detailed practical problems involved in carrying out the Survey.[5] A general criticism noted was that surveyors appeared to be classifying farmers as 'A' rather than 'B' or 'C' because they did not wish to have to describe the farmer's 'personal failings'. In addition, an 'A' or 'B' classification meant that additional visits to the farm were not necessary. The Advisory Centres were also concerned that some of the surveyors were land agents who were surveying farms in which they had an interest. These difficulties led to extra work for the Centres: one economist, based at Seale-Hayne College in Devon, noted that the work involved in the Survey was 500% more onerous than anticipated. There were also familiar problems with farmers responding to questions in unexpected ways. One farmer was reported as filling in the "engine number" in the column for "number of tractors"; this was felt to be an "error, which might give startling statistical results". In answering the question on length of occupancy some farmers were giving answers such as "I was born here" or "since the fifteenth century".

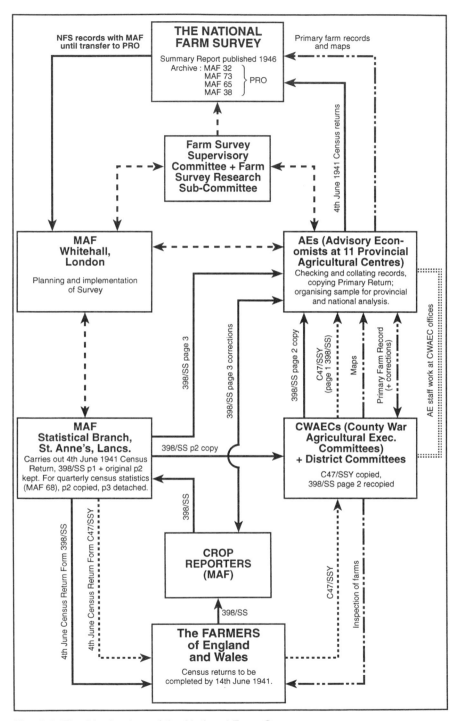

Fig. 2.1. The Mechanism of the National Farm Survey

An unexpected problem lay not so much with poorly completed forms but with the opposite: survey work that was carried out so methodically that it was well above the Ministry's stated standards. What was to be the Ministry's response to CWAEC survey officers who were eager to know the use to which their hard work, over and above the call of duty, would be put? The statistical officer for Buckinghamshire WAEC, F. W. Bateson, Oxford don and Fabian, wrote to Mr Enfield at the Ministry in January 1942 "What is really worrying me is the possibility that much of our carefully collected Bucks material may never be used. I expect you know that in some respects our original Survey Card was much fuller than your form. One of the points I have insisted on throughout is that every single field on each farm must be inspected, described and graded. We also supplement your form on such matters as rotation of crops, the ages and occupations of farm workers, the areas and lessors of grass keep land, the location of rabbit warrens and of course milk yields."[6]

Because of its complex organisation there were also significant structural problems with the Survey. There were differences of opinion between the CWAECs and the MAF Statistics Branch on what constituted a holding. This issue was complicated by the lapse of time between the 4 June returns and the date of the primary return. The main problem, however, lay in matching 4 June returns with NFS primary returns in the case of amalgamated holdings. It had been decided to exclude all holdings less than 2 ha (5 acres) from the NFS. This meant that in many cases, farms made up of an aggregate of individual dispersed holdings of less than 2 ha (5 acres) might only be partially recorded, or be missed by the NFS altogether. A re-interpretation of the 'less than 2 ha (5 acres)' rule by the Ministry led to the need to establish the association between many scattered holdings. Differences of opinion on this by the five or six people involved in making up the farm record of each such holding meant that it was certain that uniformity of interpretation in this regard would not be obtained for the whole country, or even for individual districts.

Our analysis of committee minutes, reports and papers associated with the NFS material has allowed us to reconstruct the way the Survey was carried out and how it was staffed. The complex linkages between the farmers and landowners, the surveyors (who were frequently land agents or farmers themselves), the professional academic agricultural economists and the Ministry bureaucrats have been explored. This has enabled us to develop an understanding of the development of agricultural and rural policy in the context of the prevailing social and economic attitudes and policies.

The National Farm Survey Data

To investigate the range and quality of the data provided by the NFS we analysed the nature of the responses required on the forms, how entries actually varied, and how they were either misunderstood, mistaken, amplified or subverted by the surveyors, copiers and farmers who were responsible for completing the forms. We also assessed the internal consistency of the data to gain an understanding of its quality and potential uses. We transcribed about 1% of the records composing the NFS into three databases consisting of two regional samples and one national sample. This entailed transcribing all the data from the NFS for about 3,000 holdings. The two regional samples were based in Sussex and the Midlands. For the national sample we selected one parish from every county in England and Wales. The choice of samples allowed us to examine the nature of the survey material, its collection, the relationships between different elements of the NFS and its usability in differing regional contexts, as well as the variations in the quality and nature of the data across England and Wales.

Some general items of concern which have became apparent in our transcription of data are that:

a. Some farms do not have code numbers.
b. The Census Returns for some farms are present without the Primary Return.
c. There are regional and local inconsistencies in the completion of the forms.
d. It is impossible to tell what corrections were made by the Advisory Economists.
e. A proportion of records are missing, ambiguous, illegible or demonstrably incorrect.
f. There are several internal inconsistencies within the data, such as the area of holding as found on the Primary Return and the Census Return.

The NFS provides an extraordinary wealth of data which can be used to answer a wide range of research questions, but any researcher has to come to grips with the nature of the data themselves and the degree of certainty which can be placed in conclusions drawn from them. The limitations of the NFS as an historical source are quite substantial, but without it there would simply be no information with this coverage, detail and richness. The NFS should not be treated merely as a reliable source of empirical data from which unproblematic conclusions can be drawn about the actual nature of agriculture at the time. The NFS tells us as much about the process of state intervention in agriculture and the place of farming within the national and the civil service imaginations as it does about the phenomena which were the object of its analysis. The information collected is a reflection of the system of administration used to implement it and the culture of the civil service at the time. The NFS is also very illuminating about the theoretical and methodological ideas which were being developed to deal with large bodies of data and which went on to be of importance in the ensuing development of quantitative methods in a variety of disciplines, including geography.

The Maps

Evidence from minutes and correspondence suggests that the creation of the farm maps was one of the most troublesome aspects of the Survey. Many of the base maps were out of date and did not show recent building development. CWAECs found the mapping of small holdings particularly difficult, and one minute notes that the pattern was such a 'complicated mosaic' that the final maps would serve little purpose. The Ministry's planning branch did not agree, however, and stated that in and around urban areas the 'complicated mosaic' was exactly what was required. At a meeting of the Farm Survey Committee on 12 July 1943 the view was expressed that farm boundaries shown on the maps were getting out of date and that further mapping should be postponed until after the war. However, it was decided that although imperfect, the maps would be invaluable in future planning to prevent development from cutting across the boundaries of efficient farm units.[7]

The Primary Returns and Census Returns which make up the bulk of the NFS are by themselves an enormously valuable source for the study of agriculture and rural life in mid-twentieth century England and Wales. It is, however, the survival of the associated farm maps which makes the NFS unique. No other national survey is comparable in its provision of detailed agricultural and social information which can be tied down to specific farms. MAF 73 in the PRO is a class which consists of folders containing either the Ordnance Survey 1:2,500 (12.5") sheets photographically reduced to 1:5,000 (25"), or 1:10,560 (6") sheets. A grid of 16 rectangles is stamped on the inside cover of the folders with crosses in a rectangle to indicate if any sheets are 'Wanting'. Each sheet was meant to have the OS Parcel numbers and areas transcribed onto it if they were not already present on the base sheet.

Preliminary research had shown that the completion of the maps was highly variable (Short and Watkins, 1994, pp. 290-291). Following an initial examination of a range of maps, in conjunction with the documents in MAF 32, several categories were devised to assess the quality and completeness of the maps. The map folder(s) and sheets for each sample parish were examined. This enabled us to geo-reference each holding (see below) and to assess the nature and quality of the maps at a regional and national scale. The main variables collected were the scale, edition and condition of the map sheets; the type of boundaries used to show farm boundaries; the use of shading and special keys; the system of farm referencing; the use of Ordnance Survey parcel numbers; the date the maps were drawn; whether the map was a copy or original; the authorship of the map; the use of special stamps; the treatment of non-farming land, including military land; the representation of land used by the War Agricultural Executive Committees; the treatment of fragmented holdings; and marginalia and other miscellaneous information.

Our analysis of the NFS maps confirms that they form an extraordinarily important repository of information about rural England and Wales in the mid-twentieth century. Like most important sources they are flawed. Although coverage is national, our sample suggests that around 10% of the maps are missing and some counties are more likely to have missing maps than others. A few of the maps are badly damaged, but the overall condition of the sheets is good, with most farm boundaries being clearly legible. One of the surprising results of our analysis was the quantity of information, agricultural and non-agricultural, which could be found on the maps. The various annotations and marginal comments certainly add to the value of the maps as an historical source and can provide a unique insight into land use change in the years immediately preceding the carrying out of the NFS. Our analysis showed that it was feasible to provide a geo-reference for over three-quarters of the holdings. The application of geographical information systems (GIS) is likely to prove an important means of dealing with the huge amount of data in the NFS archive, and also in linking this data to other information sources to allow for the analysis of changes over time and space. Finally, our assessment shows that the farm boundary information on the maps gives an insight into the nature of farm layout, structure and ownership in the mid-twentieth century which will prove to be a benchmark for much future research.

Conclusions

The NFS can only be understood when it is placed within the wider issue of wartime food production and the stresses imposed on farming communities. The wide spatial variations in the quality of the documentation argue for a more detailed local approach which can profitably also be located within the prevailing social, political and economic circumstances of particular localities. Such an approach would then complement the national sample studied in our project. The records will in fact shed important light on the wartime farming complexes which were moving ever faster into a 'productivist regime' (Marsden *et al.*, 1993).

State control via the CWAECs was, of course, political anathema to many farmers but the price stability and the guaranteed livelihoods now being promised were seen, even by Conservatives, to be ineluctably associated with such intervention (Bateson, 1946). The extent of such control was a subject, of course, for great debate but there can be little argument that although by 1939 the decision had been taken that stability and a permanent agricultural policy must be sought by state intervention, the next six years put in place a network of control mechanisms at local and national scales that effectively became the apparatus for a permanent post-war policy. The paper plans of Whitehall were translated into real action on some 300,000 farms, wartime practices became institutionalised and habits formed quickly (Winnifrith, 1962, p. 27; Smith, 1990, p. 88).

The NFS undoubtedly helped in the wartime construction of governmental knowledge. Through the NFS, 'local knowledges' were made available within national networks of surveillance (Matless, 1992). F. W. Bateson felt that on the whole it was the officials on the CWAECs who had taken the initiative in the collection of knowledge and in its subsequent application to change farming in England:

> "To some of these men the war gave opportunities that they never had before. The Executive Officers, Deputy Executive Officers and District Officers of the active Committees have been given a reasonable degree of power and responsibility. It is a fact that the officials have seized their opportunities with two hands, whereas on the whole the Committee members have not. Without accepting the entire gospel of the 'managerial revolution', it is clear that the WAECs have proved a fruitful breeding-ground for the new type of civil servant - energetic, self-confident, full of a special sense of public responsibility". (Bateson, 1946, p. 164)

The officials, the career civil servants and local government professionals, now ensured that agriculture became more than just a local phenomenon. It became a 'national farm' with individual farmers united through the collection and analysis of statistics on the part of government, and rendered highly visible through the newsprint and film media (Thorpe and Pronay, 1980). Its national importance in wartime and in the difficult post-war years was acknowledged. And indeed, such was the demand for statistical information in dealing with the complexities of British farming that the ministry had probably the largest statistical organisation of any department of a comparable size by the early 1960s. The NFS seen in these terms was one device in the creation of a 'fictive space' known as the 'national farm' (Murdoch and Ward, 1997). Entry into the CAP has increased demands for monitoring and control over quotas, payments and regulations, exemplified perhaps by the Integrated Admission and Control System (IACS) introduced in 1993 throughout the EU (Haines-Young and Watkins, 1996).

The release of the NFS archive means that researchers in England and Wales now have a data source for the 1940s which is comparable to the cadastral surveys available on the Continent. In its detail and coverage the NFS has no precise equal. It is less comprehensive in its spatial coverage than the Lloyd George 'Domesday' of 1910, since the latter covered all hereditaments, urban and rural irrespective of size, but the quality of information is incomparably greater (Short, 1989). And it is of more widespread relevance and has more information than the tithe surveys of the 1840s (Kain and Prince, 1985, pp. 112-113; Short, 1997). However, where the information is extant for all three dates, we now have bench marks for the 1840s, 1910 and 1940s against which to measure many aspects of farming change, not the least of which will be detailed studies of landownership and farming structure.

Acknowledgements

This research was supported by the Economic and Social Research Council, Grant R00023 5259. The authors would like to thank their co-researchers William Foot and Phil Kinsman.

Notes

1. Public Record Office (PRO) MAF 38/206.
2. PRO MAF 38/216; 38/207 (26 July 1943).
3. PRO MAF 32/208.
4. PRO MAF 38/210 (10) An early draft of the Circular is in PRO 38/207 (6).
5. PRO MAF 38/214.
6. PRO MAF 38/205.
7. PRO MAF 38/212 (17)

Chapter 3

Geographical Transition from Productivism to Post-productivism: Agricultural Production in England and Wales 1950s to 1990s

Nigel Walford

Introduction

For many years the characteristic pattern of agricultural production in England and Wales has seen arable and cash crop farming dominating the lowlands of the South and East and livestock rearing in the North and West. Inevitably there are localities which contradict such broad generalisations, for example with potato and vegetable production in Cornwall and West Wales, and dairying and sheep in areas of the South-East. Since Coppock's *Agricultural Atlas of England and Wales* was published in the 1960s (Coppock, 1964), there has been some interest shown in the interdependence between crop and livestock enterprises, with analyses undertaken to identify the mix of agricultural production systems and to generate type of farming maps (e.g. HMSO, 1967). Similarly multivariate analyses have been used to examine the interplay between cropping and livestock production enterprises in order to develop a more holistic approach to describing agricultural systems (Robinson, 1990). Local research has complemented national studies, illustrated, for example, by the case of Ilbery's work on type of farming regions in Dorset (Ilbery, 1981).

Patterns of agricultural production have received less attention from researchers in recent years as the focus has shifted towards the processes of change. Recent research has attempted to elicit and explain the motives that underpin farmers' decision-making behaviour and in particular their response to structural and policy change (Evans and Ilbery, 1989; Robinson, 1994; Gilg and Battershill, 1997). External and internal factors affecting the management of an individual farm are regarded either as influentially guiding or deterministically controlling the farm operator's actions. The contemporary approach, which seeks to understand the operation of agricultural businesses, has relegated patterns of production to the role of outcomes from a series of interconnected processes. This change of emphasis has enabled researchers to be more applied in their interpretation of farmers' responses to the economic and policy turmoil of the 1980s and 1990s.

The agricultural industries of many western societies have been beset by a 'farming crisis' in recent decades (Marsden *et al.*, 1986; Whatmore *et al.*, 1987; Hawkins *et al.*, 1993; Shucksmith, 1993) and farmers have reoriented their activities to reflect these developments. The changes have been characterised as a shift from a productivist to a post-productivist philosophy (Ilbery and Bowler, 1998). The period from the end of World War II to the mid-1980s has been referred to as the productivist era, in which a protective agricultural policy at both European and national levels cushioned farmers against the uncertainty. This helped to create conditions in which continued expansion of production was the accepted norm and common goal. Over 30 years of concerted effort by policy-makers and farmers had produced, by the mid-1980s, an economically efficient and streamlined agricultural industry in Britain and certain other western economies. A phased and initially piecemeal adjustment of agricultural policy began in the mid-1980s, with the European Union (formerly European Community) introducing various measures, including milk production quotas in 1984 and a general reform of the Common Agricultural Policy in 1992. This adjustment has sought to reduce dependence on traditional outputs, to encourage diversification and to foster more environmentally friendly farming.

With hindsight, it is surprising that the pursuit of continued growth, which dominated the post-World War II era, should ever have been viewed as indefinitely sustainable. The outcomes were structurally generated surpluses and over-production, reduced farm incomes and escalating costs of price support, with compensation for these disbenefits in the form of support for agriculturally marginal regions. In addition, modern agriculture came to be viewed as detrimental to the rural environment and farmers were castigated as the uncaring desecrators of the countryside (Shoard, 1980). It is already apparent that the supposedly dominant traditional farm, whether managed as a family or incorporated business, is undergoing change (Gasson, 1988; Arkleton Trust, 1989; Hawkins *et al.*, 1993; Bateman and Ray, 1994). It is less certain whether the spatial distribution of the new forms of farm-based economic activity will necessitate a redrafting of the agricultural map. The main aim of this chapter is to take a step towards answering this question by determining the extent to which the 'traditional' pattern of agricultural production in England and Wales has survived the transition from productivist to post-productivist agriculture. The purpose is to examine whether, as yet, there is a geographical expression of this transition.

Post-World War II Agricultural Policies and Processes

The predominance of pastoral and livestock farming in the North and West of England and Wales has been typically associated with the combination of higher rainfall, poorer soils and generally higher altitude, which favour these types of agricultural enterprise and land use. Conversely, the East and South of England, with gentler relief, more fertile soils and lower rainfall produces a comparative advantage in respect of cash crop production, despite cooler winter temperatures. Fundamental changes in the nature and structure of the industry in the post World War II period reinforced this naturally engendered division of agricultural production systems. These changes were referred to as the 'second agricultural revolution' (Bowler, 1985) implying not only that they represented a break from the past, but also that they were good for farming.

These developments were focused not only on the internal environment of the farms themselves, but also on the public and private sectors, which constitute an external environment within which management decisions are taken. Government policy and financial support underpinned these changes, which had the aim of producing an efficient agricultural industry capable of increasing output at the same time as selling food at low prices. A range of measures were used to implement these policies and provided public financial assistance to the industry. These included product marketing boards, regulation of farm workers' pay through standardised wage rates, maintenance of prices through an annual review system, tax allowances for capital investment and direct subsidy of farm improvements (e.g. drainage works). The 'public purse' also funded consultancy, training and research through the Agricultural Development and Advisory Service, the county agricultural colleges and the universities. The external private sector helped to foster the growing commercialisation and globalisation of the agro-food industry (Whatmore, 1995; Robinson, 1997). New companies emerged and existing ones expanded with the specific purpose of supplying farmers with various goods and services. This was part of a process of externalisation with respect to the inputs and outputs of agricultural production processes. Collectively these developments can be regarded as the definitive components of the productivist phase in agriculture. Many farmers, influenced by external pressure and subsidy, and by their self-motivated pursuit of profit, contributed uncaringly, if not unwittingly, towards surplus production.

From a spatial perspective, the related processes of concentration, intensification and specialisation (Ilbery and Bowler, 1998) enhanced the pastoral-cropping divide in the agricultural geography of England and Wales. These processes were grounded in a rational and scientific approach to production, in which it is believed that farmers engage in the most efficient (economic) enterprises to an optimum level. Farm resources would thereby be utilised efficiently and would not be dispersed across a range of activities. Specialisation is a farm-centred process, paradoxically involving both simplification and complication of a farm production system. A simplified, more focused system might arise from the removal of some enterprises and the expansion of others. Added complexity can result from the higher risk and skill requirements associated with dependency upon the success of a narrower range of products. The decision to implement this type of restructuring is usually justified in economic terms, and centres on the notion of farmers concentrating their investment and entrepreneurial effort. A number of factors may be connected with the decision to specialise in this way. Investment in higher capacity machinery may be possible when financial resources are more concentrated. At a time of increasing sophistication in production methods, technical expertise may be developed in the agronomic and marketing aspects of a more limited range of enterprises.

Concentration of production within agriculture is essentially a spatial process operating at local or regional scales. It can simply be viewed as the aggregate outcome of the collective behaviour of individual producers who are specialising their farm businesses. Thus, if farmers in a region independently or collaboratively adjust their production systems in the same way, spatial concentration on this enterprise will ensue. External stimuli can enhance the tendency towards spatial concentration. For example there has been a progressive reduction in the number of sugar refineries in England and Wales over several decades, which has resulted in the cultivation of sugar beet becoming more spatially concentrated to within convenient drive times of the factories used to process this bulky crop. The opening or expansion of plants can have a similar effect, with freezing and canning factories providing an outlet for horticultural products. The direction of any cause and effect relationship may be unclear, although some form of association is apparent.

Measurement and Methods

The principal source of data on agricultural production in England and Wales is the annual Agricultural Census (Clark, 1982), from which statistics on the areas of crops and numbers of livestock are published (e.g. Ministry of Agriculture Fisheries and Food, 1997). Year-on-year changes in these statistics should be interpreted cautiously. For this reason periodic averages have been used, which smooth the effect of annual, possibly untypical fluctuations, and highlight underlying trends. Three-year averages at 10-year intervals for the counties of England and Wales for 1954-1956, 1964-1966, 1974-1976, 1984-1986 and 1993-1995 have been used. The immediate post-war period has been omitted in order to avoid any residual effect of the wartime 'plough-up' campaign, the presence of the Women's Land Army and other emergency measures (see Ch. 2).

There are two further problems when attempting to analyse spatio-temporal change. First, the boundaries of the aggregation units may not remain constant. Counties in England and Wales experience minor boundary adjustments on an intermittent basis. However, more serious discontinuities have arisen during periods of local government reorganisation in 1974 and 1996-1998. For this reason the years 1993-95, rather than 1994-96, have been used as the most recent period, since changes in the structure of administrative areas precludes published Agricultural Census statistics for 1996 from being combined with those of 1994 and 1995. The second problem is that, irrespective of change in external boundaries, differences in the proportion of agricultural and non-agricultural land within a county can also occur. Figure 3.1 illustrates the problem by depicting two hypothetical counties, which are small and large in their total area. Parts A and B illustrate that, despite having different total land areas, the relative proportions of the three types of land use are the same, although the absolute areas are different. Parts B and C illustrate how the relative proportions of the land uses can change over time (non-agricultural and fodder areas have both decreased). Such changes may occur to the extent that the absolute areas of the agricultural land and of its constituent categories are identical to those found in the small county overall. A change from a small to a large area of a particular type of agricultural land use may be interpreted as indicating greater concentration of that enterprise. However, the situations depicted in Fig. 3.1 highlight that some care is required before reaching such a conclusion. Similar issues would arise if the land use change analysis was carried out at a different geographical scale, for example districts or parishes.

The various types of land-based agricultural production have been divided into two broad groups, cash and fodder cropping. Cash cropping refers to those arable enterprises whose products are typically sold off the farm, including cereals (wheat, barley, oats and rye), other field-scale crops (potatoes, oil seed rape and sugar beet) and horticultural crops. Fodder crops can act as a surrogate indicator of livestock production and included rough grazing, permanent pasture, temporary grass and other forage (maize, root crops, etc.). This division is not perfect in forming a separation between cropping and livestock production, since some cereals grown on farms may be retained as feed for livestock and some fodder crops may be produced for sale.

These data problems have been addressed in this analysis by standardising the agricultural statistics through conversion of the raw figures to Z scores and to location quotients (LQ). Standardisation minimises any distortion arising from changes to county boundaries and from differing proportions of agricultural land.

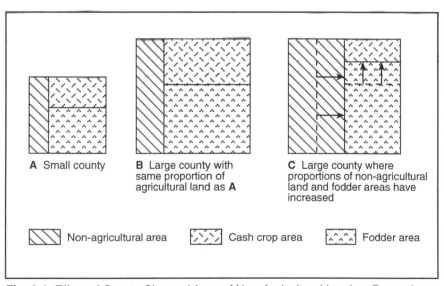

Fig. 3.1. Effect of County Size and Area of Non-Agricultural Land on Proportions of Cash and Fodder Cropping

Both procedures indicate which counties have relatively more or less of a particular type of agricultural production, but the former relates this to the normal probability distribution. Prior to standardising as Z scores, the areas of the two categories of agricultural land were expressed as proportions of total land area per county. Each county's Z score reflects its position with respect to the mean for the variable concerned. Thus counties with extreme negative or positive values have respectively very low or very high areas of the particular crop. A Z score greater than or equal to $+/- 1.96$ indicates that the county in question lies in one of the tails of the normal distribution and is significantly different from the 'average' situation. A LQ can be interpreted in a similar way and is a measure of the extent to which individual members of a group of spatial units contain more or less of a particular phenomenon in comparison with a norm, such as the population mean. LQs in excess of 1.0 signify relative concentration of the characteristic, whereas values between 0.0 (the minimum) and 1.0 denote the reverse.

Changing Patterns of Agricultural Land Use

The cash and fodder crop areas expressed as proportions of total area and converted into Z scores are given in Fig. 3.2 for each of the decades under consideration. As a general guideline, the shorter the bars the closer the county is to the national situation. Those counties with a high positive Z score for cash cropping and a high negative score for fodder signify a concentration on crop production and low emphasis on livestock. The converse indicates concentration on livestock enterprises associated with forage land and relatively minor cultivation of cash crops. The extremes are represented by a relatively small number of counties in each case where either cash cropping or livestock enterprises dominate: the remaining counties have mixed agricultural land use and fall into three sub-groups (see right-hand columns of Table 3.1). The distribution of counties in Fig. 3.2 appears to confirm the traditional livestock and cash cropping division. Some counties have changed positions over time, whereas for others there is evidence of tenacity and strengthening in the emphasis on their dominant type of agricultural land use.

The geographical distribution of cash cropping and fodder land is examined in more detail in Fig. 3.3. The series of maps presents the difference between the proportions of cash and fodder crops in relation to total land area, which has been mapped according to units of the standard deviation either side of the mean difference. If cash crop and fodder land occurred in equal proportions in a given county, the difference would be zero $(0.5 - 0.5)$. However, when the cash cropping dominates, a high positive difference is obtained (e.g. Norfolk and Lincolnshire scored $+0.42$ and $+0.38$ respectively in the mid-1980s). In contrast, counties with high negative values have larger proportions of land under fodder crops, for example in the mid-1980s Cornwall, Cumbria and Dyfed respectively recorded -0.52, -0.60 and -0.80. The cash-cropping region spread out from the core area of Cambridgeshire in the 1950s to absorb Huntingdon and Peterborough in the 1960s and then Lincolnshire in the 1970s. By the mid-1980s and continuing into the 1990s the remaining counties in eastern England were included in this intensely arable region. However, the significant change as post-productivism has dawned is that even in these counties the difference between the proportions of cash and fodder land in relation to the total area has fallen. The mean difference having progressively reduced from the 1950s to the 1980s (-0.37 to -0.07), had reversed direction by the 1990s (-0.12). This suggests a modest return to a more balanced division between cash cropping and fodder land. The concentration on fodder cropping in the counties of central and northern Wales is evident throughout the period. These have been joined by western counties in England in different decades: Cumberland and Westmorland (1960s); Cumbria, Devon, Northumberland and Somerset (1970s); Avon, Cornwall, Devon and Somerset (1980s); and Cumbria, Devon and Lancashire (1990s).

Table 3.1. Types of county based on Z scores of cash crops and fodder area proportions

County Type	Cash Crop	Livestock	Mixed	Mixed	Mixed
Sub Type			Fodder Dominant	Cash Crops Dominant	Cash Crops nor Fodder Dominant
Cash crop Z score	High positive	Moderate negative	Low negative	Moderate positive	Low/ moderate same sign as Fodder
Fodder Z score	Moderate negative	High positive	Moderate positive	Low negative	Low/ moderate same sign as Cash Crops

Table 3.2 lists those counties at the upper and lower extremes in terms of the difference in the proportion of cash and fodder cropping in relation to total land area. Many of the counties feature in the top and bottom five of the ranking in most decades. In counties where the positive proportionate difference increased (e.g. Norfolk), there was growing concentration on cash crops, conversely a series of decreasing negative values (e.g. Caernarvonshire and its successor county Gwynedd) signified expansion of fodder cropping. The increasing value of the proportionate difference is clearly evident with respect to the top five cash cropping counties up to the 1980s, which is mirrored by rather less dramatic decreases in those counties dominated by fodder cropping. By the 1990s there is evidence of a reversal having occurred, for example with Lincolnshire falling from 0.63 to 0.39 and Gwynedd rising from –0.70 to –0.75.

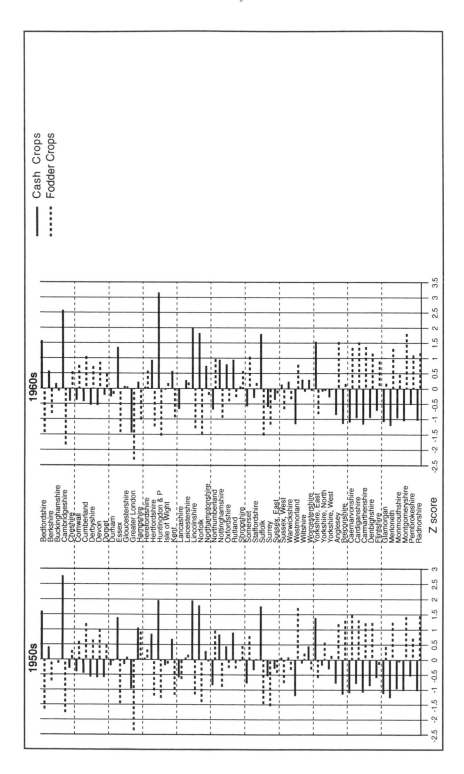

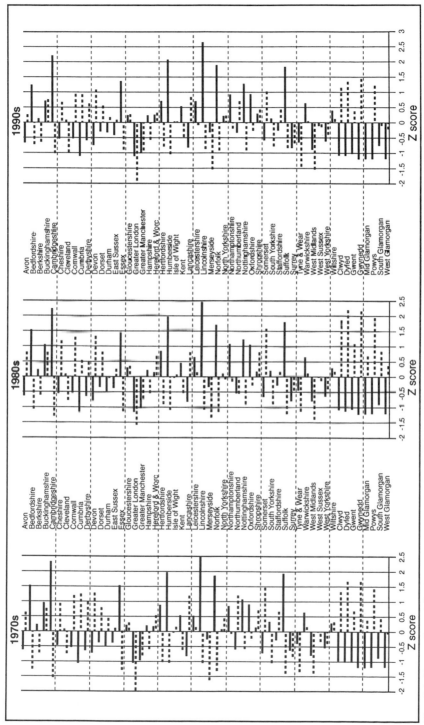

Fig. 3.2. Proportion of Cash and Fodder Crop Areas In Relation to Total Land in England and Wales as Z Scores (1950s to 1990s)

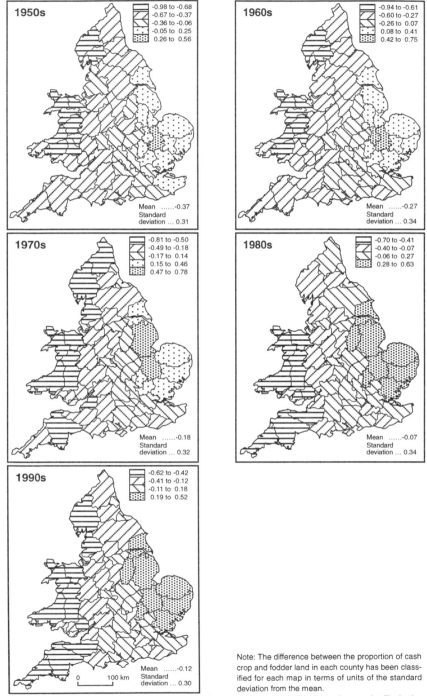

Fig. 3.3. Difference in Proportions of Cash Crops and Fodder Areas in Relation to Total Land in the Counties of England and Wales (1950s to 1990s)

Note: The difference between the proportion of cash crop and fodder land in each county has been classified for each map in terms of units of the standard deviation from the mean.

Table 3.2. Extremes of cash and fodder cropping in counties of England and Wales, 1950s to 1990s

1954-6	1964-6	1974-6	1984-6	1993-5
Top Cash Cropping Counties				
Cambridge-shire (0.39)	Huntingdon & Peterborough (0.58)	Cambridge-shire (0.52)	Lincolnshire (0.63)	Lincolnshire (0.52)
Bedfordshire (0.19)	Cambridge-shire (0.58)	Lincolnshire (0.47)	Cambridge-shire (0.62)	Cambridge-shire (0.46)
Suffolk (0.19)	Lincolnshire (0.33)	Suffolk (0.42)	Humberside (0.54)	Humberside (0.45)
Huntingdon & Peterborough (0.18)	Suffolk (0.33)	Norfolk (0.40)	Suffolk (0.52)	Norfolk (0.40)
Norfolk (0.18)	Norfolk (0.33)	Humberside (0.37)	Norfolk (0.49)	Suffolk (0.40)
Top Fodder Cropping Counties				
Westmorland (-0.87)	Montgomery (-0.78)	Gwynedd (-0.69)	Gwynedd (-0.70)	Dyfed (-0.62)
Caernarfon-shire (-0.81)	Carmarthen-shire (-0.73)	Dyfed (-0.66)	Dyfed (-0.68)	Powys (-0.58)
Radnorshire (-0.79)	Merionethshire (-0.73)	Powys (-0.65)	Powys (-0.65)	Clwyd (-0.58)
Montgomery-shire (-0.79)	Cardiganshire (-0.72)	Clwyd (-0.60)	Clwyd (-0.62)	Gwynedd (-0.57)
Merionethshire (-0.78)	Caernarfon-shire (-0.71)	Cumbria (-0.60)	Somerset (-0.48)	Cumbria (-0.51)

Notes: The figures in brackets are the difference between the proportions of cash and fodder cropping land in relation to total area in each county. High positive values indicate a concentration of cash cropping, whereas high negative values denote the dominance of fodder crops.
Source: United Kingdom Agricultural Census

An alternative view on the changing geographical pattern of agricultural production is given by the series of LQ maps in Fig. 3.4. The three class intervals for each agricultural land use have been formed by making a central class composed of the top 25% of the counties in the range below 1.00 and bottom 25% in the range above 1.00. Many counties move from the low class for cash cropping into the high class for fodder and *vice versa*, with a central group remaining in the middle group for both land-uses (crossed shading in Fig 3.4). In many respects the distributions depicted here reinforce the pattern of proportionate differences in relation to total land area. There is some indication that the area concentrating on cash cropping has fluctuated throughout the period, with counties such as Berkshire and Hampshire moving in and out of the group. The range of LQs for cash cropping tended to be larger than for fodder, perhaps indicating a greater tendency towards extremes of production in the former case. For example, the top cash cropping county in the 1950s, Cambridgeshire with 2.65, and in the 1990s,

Lincolnshire with 2.12, compare with the top fodder county in the 1950s, Westmorland with 1.34, and in the 1990s, Gwynedd with 1.60. At the opposite end of the ranges, Cambridgeshire had a fodder LQ of 0.37 in the 1950s Merionethshire a cash cropping LQ of 0.09, similarly in the 1990s Lincolnshire achieved a fodder LQ of 0.29 whereas Gwynedd only managed 0.05 for cash cropping. In other words, those counties with high concentrations of fodder crops had extremely low areas of cash cropping, whereas the cash cropping counties had a modest level of fodder production.

Comparison of Figs 3.3 and 3.4 clearly reveals that the cash cropping Eastern Counties and the fodder growing (livestock) western ones are separated by a group of less clearly defined counties. Here the data values in Fig. 3.3 lie between -1 and +1 standard deviations of the mean and the LQs for both types of agricultural land use tend to place the counties in the central group either side of 1.0. The counties with a strong emphasis on fodder cropping had rising LQs up to the 1980s, but the maximum reduced slightly in the 1990s. The maxima LQs for cash cropping displayed the opposite trend with a rise from 2.01 to 2.12 between the 1980s and 1990s.

Discussion and Conclusions

The characteristic geographical division of agricultural land use in England and Wales has been maintained and strengthened up to the mid-1980s. The analysis suggests that traditional forms of agricultural activity in the arable heartland have intensified over the post war period. The fodder (livestock) area experienced some enlargement and fluctuation throughout the period. During the transition to post-productivism, there is some indication that, although the cash cropping and livestock division is still in place, the intensity of each activity has reduced. The lower proportionate differences between cash crop and fodder areas in relation to total land area may indicate some extensification of production, although the traditional areas of concentration persist. The LQ analysis perhaps suggests that cash cropping has become more concentrated in its traditional heartland. From a policy perspective, the lower figures possibly reflect the effect of set aside measures requiring farmers to take arable land out of production. The central swathe of counties with mixed agricultural production, which separates the concentrations of cash crop and fodder production, constitutes the area of England and Wales, particularly in the Midlands and South-East, where higher proportions of non-agricultural land are found. This area is also likely to have experienced greater pressure for development, since it incorporates the major urban centres. This mixed cash cropping and livestock group also divides reasonably clearly between those of the North and West where fodder tends to dominate and those of the South and East where cash cropping is more important. Thus, even within this mixed group the overall national division is evident.

The analysis presented here has only considered two broad groups of land-based agricultural enterprises, clearly the more detailed changes in individual cash crop and fodder enterprises as well as comparable analysis with livestock numbers

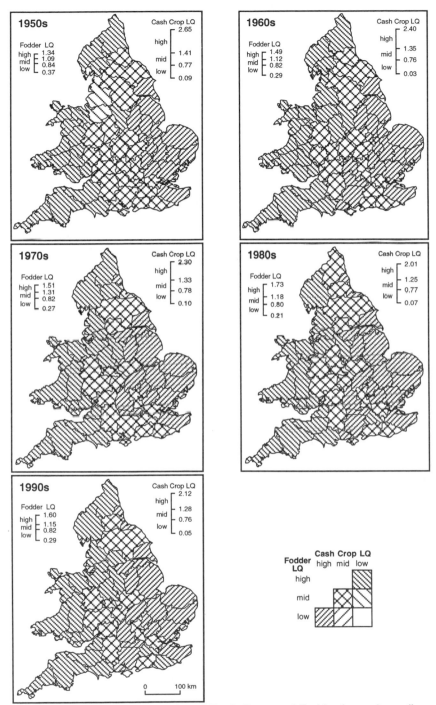

Fig. 3.4. Changing Concentration of Cash Crops and Fodder Areas According to Location Quotients in the Counties of England and Wales (1950s to 1990s)

would provide a more comprehensive picture. The spatial scale of the analysis could also be reduced to districts or perhaps even parishes, although both of these have their attendant problems (Clark, 1982). The spatial scale chosen here allows examination of the broad changes in agricultural land use, however a finer scale of analysis would provide more insight into local variations. The emphasis in this chapter has been on the extent to which the transition from productivism to post-productivism has disturbed or re-enforced the traditional patterns of agricultural production in England and Wales. It would appear that under the 'new regime' of post-productivism the processes of specialisation and concentration remain in evidence, but perhaps on a lesser scale.

The restructuring of the agricultural industry and the geographical manifestation of this process are the outcome of decisions taken by farmers and their families to alter their income-generating activities and the mix of commercial enterprises. Clearly these decisions are made within a political, economic and social framework which guides or entices farmers towards certain ends. Although only imperfect knowledge exists about the full extent to which the farming community has restructured and diversified, there appears to be sufficient evidence to suggest that the conventional definition of a farm needs some revision. The traditional conceptualisation of a farm is as an economic business unit engaged in land-based enterprises concerned with the production of plant and animal products for food or industrial use. These activities remain at the heart of the work carried out on many farms, nevertheless a significant number are now also concerned with the leisure, manufacturing and retailing industries. They have ceased to be exclusively primary sector businesses and many now incorporate elements of secondary and tertiary sector activity as well. No doubt the relatively small scale, but nonetheless insidious, manner in which these changes have come about has masked a full appreciation of their impact. It is time to determine the extent to which farms have become factories, shops and recreation centres.

Chapter 4

Moralising Nature?: The National Rivers Authority and New Moral Imperatives for the Rural Environment

Susanne Seymour, Philip Lowe, Neil Ward and Judy Clark

Introduction

The management of water systems in England and Wales has undergone several reorganisations over the past 30 years but none have been more politically contentious and publicised than the 1989 changes linked to water privatisation. These developments witnessed the replacement of the former Regional Water Authorities (RWAs) by new private water companies and, after protracted deliberations, a new regulatory organisation, the National Rivers Authority (NRA). The staff and duties of the RWAs were divided between the new private water companies, which were given responsibility for drinking water supplies and waste water treatment, and the NRA, which took over the roles of river catchment management and regulation (Maloney and Richardson, 1994). From the onset, the NRA proclaimed itself as a new moral force in water regulation. Pledging to act as "a tough and effective regulator", it styled itself as "Europe's strongest environmental protection agency" (NRA, 1989a). Such claims run counter to the established British approach to environmental regulation typically characterised by informality and negotiation, with only rare reliance on legal sanctions (Hawkins, 1984; Vogel, 1986; Lowe and Flynn, 1989). Instead, the NRA endeavoured to draw more explicitly on an enhanced moral view of the environment among environmental groups and the public at large and, in particular, on heightened public concerns over water pollution in the context of privatisation. In practical terms it tried to accommodate environmental interests within its organisational structure, to project a tough regulatory approach, to establish national standards of regulation and ensure they were consistently applied across the regions and to encourage public participation in the reporting of suspected pollution. On 1 April 1996 the NRA was incorporated into the Environment Agency along with Her Majesty's Inspectorate of Pollution (HMIP), the Waste Regulatory Authorities and some small technical units of the Department of the Environment (DoE) (from 1997 the Department of Environment, Transport and the Regions – DETR) dealing with waste and contaminated land. The Environment Agency created a cross-source agency which helped fulfil the UK government goals of integrated pollution control (DoE, 1990; Environment Agency, 1996a).

This paper uses concepts of social constructionism (Hannigan, 1995) and policy communities (Marsh and Rhodes, 1992) to assess the NRA's role in the 'moralisation of nature', in particular through its institutional and operational changes and its regulation of agriculture. The process of 'environmental moralisation' we are considering involves a number of aspects. Firstly, this idea of moralisation refers to the development and institutionalisation of an ethos which draws on groups and discourses promoting concerns about the natural world and which adopt an ecocentric stance, lending nature value in its own right (O'Riordan, 1981). Accompanying this is a concern that the organisation draws on public support and is publicly accountable in its actions. A second stage in the moralisation process is for these values to become operationalised within an organisation. Under such conditions policy solutions to environmental problems shift from a technical fix approach, often using end-of-pipe solutions, to a concern with managing and regulating complete production processes. In addition policy processes are opened up to wider public involvement and scrutiny. Pollution itself is generally conceptualised less as a form of rule breaking and more as an 'environmental crime' and close attention is likely to be paid to key actors' environmental values and the influence these have on actors' practices (Weale, 1992; Lowe *et al.*, 1997). We review the NRA's claims and its operation at national, regional and local levels, considering how far environmental and public interests were accommodated during the lifetime of the Authority and the extent to which values and strategies within the organisation changed, particularly in terms of preventative policy, rates of prosecution and levels of pollution. To assess the NRA's role in the construction of environmental morality we use a case study of the regulation of pollution by dairy farms, looking specifically at its operation in the South West region. Finally, we identify a set of influences working against the establishment of this new moral force in environmental protection.

The National Rivers Authority

When the NRA was created in 1989, it inherited the bulk of its infrastructure (for example, regional boundaries) and its staff from the old RWAs. However, the Authority deliberately sought to project an image sharply different from the former RWAs, which had come to be seen as inadequate protectors of the environment, and from the newly privatised water companies about which there was considerable public suspicion. In particular, the Authority set out to align itself with the environmental movement and public concerns, and to draw on their moral authority in terms of both environmental and social legitimacy, something which the RWAs had failed to do.

An important strategy was to present itself as a brand new organisation, with novel powers, a fresh outlook and no history. An introductory leaflet, proclaiming the Authority to be *Guardians of the Water Environment*, opened with the following foreword by Lord Crickhowell, a former Conservative minister and the new chairman of the NRA:

"September 1st, 1989, represented a turning point in the history of environmental protection in England and Wales. On that date the National Rivers Authority took up its duties. Overnight the NRA became the strongest environmental protection agency in Europe."

Pressing home the point, the leaflet continued by arguing that the "existence of a powerful, impartial and independent organisation with a clear statutory responsibility to carry out its duties transforms the way in which our water environment is guarded" and that it constituted an "immense improvement on the arrangements that have existed before".

By distancing itself from its forebears and adopting a moral rhetoric, the NRA sought immediately to seize the high ground of environmental morality and to claim public support for its timely appearance on the regulatory scene. Arguing that it had come into existence "at a time when there has never been more concern at the damage mankind inflicts on the natural systems on which we all depend", the Authority highlighted that foremost among its "far-reaching responsibilities" was the duty to "control pollution and improve the quality of our country's river systems and coastal waters . . . for the sake not only of this generation - but of those to come". Claiming it would adopt a much tougher approach in regulating the water environment, it regarded this as legitimated not only by new legal powers but also by widespread public support:

"The strength of the NRA goes beyond the authority invested in it by Parliament and the assurances of the government that it will give us all the necessary support. It taps as well the vast reservoir of public opinion, with its strong feelings about the importance of the environment" (NRA, 1989a).

The new environmental morality which the Authority sought to draw on, and which informed its creation, under encouragement from the environmental movement, elevated pollution and industrial risks to the status of 'crimes' (Knowland, 1993; Lowe *et al.,* 1997). It has come about, at least in part, because people have become increasingly distanced from the causes but not the consequences of pollution and technological hazards. Shifts over the past three decades in employment structure towards service sector jobs (Savage *et al.,* 1992; Marsden *et al.,* 1993, pp. 1-40) has meant fewer people are exposed to a personal conflict between economic and environmental welfare, and environmental concerns have grown. Indeed, there has been a rapid rise in popular environmentalism. The number of environmental groups has expanded and membership of a wide range of these accelerated significantly in the 1980s in particular. For example, by 1993 Greenpeace (set up in 1971) had 410,000 UK members and a longer established group, the Royal Society for the

Protection of Birds (RSPB) reported 750,000 adult members by 1995 (rising from around 12,000 in 1960) (McCormick, 1991, pp. 34-38; Winter, 1996, pp. 186-187). Furthermore, the new NRA was presented with an opportunity to cultivate its own policy network as established policy communities were thrown into disarray at the time of water privatisation, allowing environmental interests, at least temporarily, to have a greater say (Maloney and Richardson, 1994). Faced by high levels of pollution from agriculture and a legacy of a strong agricultural policy community, albeit one under increasing challenge in the late 1980s over issues of production and environment (Cox *et al.,* 1986; Winter, 1996), it is unsurprising that the Authority sought to strengthen a looser network of environmental interests and build its constituency there.

The NRA aligned itself with popular concern for the environment through new consultation procedures and publicity, courting both environmental groups and the wider public. The consultation procedures of the Authority accommodated environmental groups in ways which those of its predecessors did not. Under the RWAs, environmental groups were not part of the water policy community. The only environmental interests represented on boards of the water authorities tended to be MAFF appointees from Regional Fisheries Committees (Lowe *et al.*, 1997, p. 63). By contrast, the NRA made an early strategic commitment to undertake its statutory liaison duties "through a much more open consultation process with . . . conservation bodies than pertained under previous arrangements" (NRA, 1989b) and brought a number of moderate conservationists onto its Board. In addition, a network of Regional Rivers Advisory Committees was established, the role of which was to comment on draft national policy as well as regional regulatory and management matters. These included among their membership representatives from a range of environmental organisations, for example Friends of the Earth (FoE), the Council for the Protection of Rural England (CPRE) and the Royal Society for Nature Conservation (RSNC), as well as more traditional industrial and agricultural interest groups such as the Confederation of British Industry (CBI), the National Farmers Union (NFU), the Country Landowners Association (CLA) and the water companies. These meetings were also open to members of the public and the press (NRA, 1990a).

The latter arrangement was just one way in which the Authority publicised its work, a strategy in which it invested considerable time and resources. Indeed, the NRA claimed that a key achievement of its first year of operation, was the establishment of "a corporate identity and strong public image . . . through effective media relations" (NRA, 1990a, p. 4). Its eye-catching logo of a leaping salmon was displayed prominently on its vans, buildings, posters and leaflets. Press releases concentrated on "publicising pollution incidents and their effects, and the NRA's regulatory role" (NRA, 1990a, p. 39) while its own newspaper projected its image as The Water Guardians. A key means of actively involving and associating the public with its work was through the reporting of pollution incidents (Ward *et al.*, 1995). Concerned to enlist the public on the ground, as active 'pollution watchdogs', from 1990 a number of NRA regions established a free, round-the-clock Pollution Hotline to encourage the public to ring in with reports of suspected pollution and in 1994 this was converted into a national service. To publicise the launch of the Hotline in the South West region, tens of thousands of specially produced cards carrying its number were distributed. Notices alerting the public to the threat to rivers from farm pollution and encouraging people to report any incidents were visible on most village notice boards and pinned to telegraph poles in even the most isolated parts of the Devon countryside. Hotline cards could be found in local libraries, police stations and even the local branch of the NFU. Clem Davies, the region's Environmental Protection Manager, summed up the strategy as follows, "Everyone has a role to play in caring for our rivers. . . . The Authority has staff in the field but the public are our ears and eyes too" (NRA, 1990b). The encouragement of people to be vigilant over pollution gave the strategy the air of a moral crusade and, in general, such public relations drew upon and encouraged expectations that the NRA would take a tough stance on pollution.

The promotion of its status as a national organisation, operating national policy standards, also helped the NRA in its attempts to establish itself as a new moral force in regulation, particularly since such claims contrasted markedly with the former regional structure of water management. Enhancement of the Authority's national status was facilitated by a number of developments, including the establishment of a sizeable national headquarters and the formation and monitoring of nationally applied policies and standards. Although initially the Authority had only a small national headquarters in London employing around 60 mainly administrative staff, in 1990 it opened a new larger headquarters at Bristol and by 1992 staff levels at headquarters had more than doubled (NRA, 1990a, p. 47; NRA, 1992a, p. 38). This growth allowed the Authority to build up a central body of expertise, particularly in relation to environmental protection, a process which facilitated both the development and implementation of more environmentally sensitive national policies. The Authority also made concerted efforts to ensure national policies were consistently applied in the regions. To facilitate and monitor this process, from 1990 NRA headquarters began issuing a series of Policy Implementation Guidelines (PIGs) and in 1991 created a new post of Director of Operations, with responsibility for overseeing the application of national standards (NRA, 1992a, p. 11).

The NRA also accommodated environmental interests by altering the policy priorities and staffing balance inherited from the RWAs. Whereas the RWAs were dominated by chemists and engineers and concentrated on water supply and sewage treatment, the NRA's priorities and its staff expertise were more tightly focused on environmental protection. The number of staff trained in environmental sciences both at its headquarters and in its force of Pollution Inspectors on the ground was increased and concern with the ecological integrity of the whole water environment increased over the lifetime of the organisation. Pollution control and the associated tasks of water quality regulation and environmental monitoring figured significantly in the Authority's responsibilities. Water quality monitoring and pollution control procedures were updated and charges introduced for discharge consents and clean-up costs following pollution incidents (NRA, 1992a). In-depth studies of the regulatory system suggest that NRA Pollution Inspectors held a more morally charged view of pollution than did similar RWA staff in the 1970s (Hawkins, 1984; Knowland, 1993).

The alterations also included a declared intention to take more legal action. The NRA's first annual report (for 1989/90) laid plans to take polluting events seriously, to pursue a higher rate of enforcement than the former water authorities had and to press for stiffer penalties following successful prosecutions (NRA, 1990a). The Authority also pledged to "enforce the pollution law even-handedly according to the scale of damage caused to the water environment" (NRA and MAFF, 1990, p. 1). One aspect of injecting a more legally oriented element into regulation, whilst at the same time standardising procedures, was for the Authority to instigate nationally co-ordinated legal training for all new Pollution Inspectors. Another was the issuing of national guidelines for prosecution decisions for the newly established (1990) categories of major, significant and minor pollution incidents.

However, the NRA's responsibilities did not extend to all aspects of water pollution. The regulation of major industrial sources was the duty of HMIP, established in 1986 and under the 1990 Environmental Protection Act responsible for integrated (or cross-media) pollution control. In addition, the 1989 Water Act gave the privatised water companies powers to regulate discharges of trade effluents to the public sewers, leaving them as the most prominent dischargers to rivers which were regulated by the NRA. Since the government relaxed discharge conditions on sewage works for a period to avoid prosecution of the newly privatised companies and as Authority staff had to reach a modus vivendi with their former colleagues, in its early years the NRA's scope to pursue its crusade against pollution was considerably circumscribed with respect to urban and industrial sources and it had to look elsewhere to prove its regulatory claims (Lowe *et al.*, 1997, pp. 82-90).

The NRA and New Approaches in the Regulation of Agricultural Pollution

Outside of urban and industrial areas, the most prevalent source of pollution encountered by the new NRA was agriculture. Farm effluents did not fall within the scope of HMIP and they were not usually discharged into the water companies' sewers. Most pollution from farm effluents occurred through direct discharge or runoff into rivers and streams which were the NRA's responsibility. Many of the staff who had made an issue of farm pollution during the last years of the RWAs - the members of the Water Authorities Association's (WAA) Farm Waste Group and their colleagues - secured senior positions in the new Authority. Almost inevitably, therefore, the control of farm pollution was central to the new Authority's efforts to make good its claim to be the guardian of the water environment. As remarked in the press, "the farming community would be one of the NRA's chief targets" (*The Financial Times*, 31 October 1990).

By the time the Authority started operating there was political recognition of water pollution from dairy farm effluents and from agricultural nitrates (Seymour *et al.,* 1992; Watson *et al.,* 1996; Lowe *et al.,* 1997, pp. 60-87). Indeed, the 1989 Water Act set out particular measures to combat both. The Ministry of Agriculture, Fisheries and Food (MAFF) was given responsibility to administer a voluntary pilot control scheme (Nitrate Sensitive Areas) aimed at reducing nitrate pollution of drinking water (a response to problems in achieving the standards of the 1980 EC Drinking Water Directive and anticipating a new Nitrate Directive setting out land use control measures). The 1989 Act gave the new NRA significantly enhanced powers to tackle agricultural pollution, including controls for the first time over the handling and storage of farm effluents. These, commonly known as the Farm Waste Regulations, were operationalised in 1991 (DoE and Welsh Office, 1991). In addition, the 1990 Environment Act raised the maximum fines that the NRA could seek to have imposed in magistrates' courts for water pollution offences from £2,000 to £20,000. When the EC Nitrate Directive was finally agreed in 1991 designation and policing responsibilities for the new compulsory Nitrate Vulnerable Zones (NVZs) were also shifted onto the NRA and subsequently the Environment Agency (Environment Agency, 1998, pp. 102-106).

The NRA and Strategic Agri-environmental Policy

The NRA was thus presented with an entrée into agricultural policy circles by the farm pollution issue and by the temporary disarray of policy communities at the time of water privatisation. Whereas the RWAs had had no standing in the agricultural policy community, the NRA was consulted closely on the Farm Waste regulations, drawn up by the DoE (its parent department), and acted as a statutory consultee in the drafting of the new *Code of Good Agricultural Practice for the Protection of Water*, issued by MAFF in 1991 (MAFF and Welsh Office, 1991). In addition, qualification for the MAFF grants for pollution control measures instigated in 1989 under its Farm and Conservation Grant Scheme (which ran until 1994) was also dependent on NRA consultation.

The RWAs had had little influence over agricultural production policy and their albeit low key efforts to control agricultural effluents had been focused almost exclusively at the farm level. The approach they adopted to farm pollution regulation had been generally dominated by an ethos of "information and persuasion" (WAA and MAFF, 1986, p. 10) and in the 1970s they were reported to believe that "negotiation rather than punitive measures, will lead to better co-operation" (Weller and Willetts, 1977, p. 29; Lowe *et al.,* 1997, pp. 53-57). Even when problems with rural water quality and fish stocks had begun to emerge in dairying areas in the early 1980s, a policy of farm visits had generally been adopted in which only those farmers who persistently ignored the RWA's warnings and advice had been in danger of prosecution (NRA, 1992b; Lowe *et al.,* 1997, pp. 90-91). By contrast, the NRA was keen to take a more active stance in strategic agricultural policy and to develop a regulatory approach in which recourse to legal procedures figured more strongly.

Thus, in its early years, it became clear that the NRA had ambitions to challenge agricultural production and policy decisions. The Authority was openly critical of policies that had led to the intensification of farming without having due regard for the environmental consequences, particularly those relating to livestock production. In 1990, the chairman of the Authority declared publicly that "in many areas of intensive livestock production, present levels of stocking and slurry disposal are damaging the environment" (*The Financial Times*, 31 October 1990). Within the NRA there was a strong feeling that problems with livestock effluents were endemic in certain regions (including the South West) with many of the dairy farms overstocked, in the wrong place or with production systems that were intrinsically risky. The threat of prosecution of individual farmers and investment in pollution control facilities dealt with the symptoms of the problem but the fault was seen to lie more fundamentally with agricultural policy for having encouraged the intensification of dairy farming. The NRA thus saw it as much more important to seek the redirection of farming practices and policies towards ones that posed reduced environmental risks and pressures.

However, despite the legislative changes, the NRA experienced difficulty in influencing agricultural production policy, laid down at EU level through the Common Agricultural Policy (CAP) and at national level by MAFF. The Authority's sponsor department in government was the Department of the Environment and although there were policy links with MAFF, these mainly related to its flood defence and fisheries duties, which were grant-aided by the Agriculture Ministry (NRA, 1990c). Despite the agricultural policy community suffering a series of threats since the 1970s, productivist interests and technical fix solutions to environmental problems continued to hold firm in MAFF. While the Ministry was willing to co-operate with the NRA over pollution control grants, its technical fix mentality dominated the strategy (Lowe *et al.*, 1992) and there was more resistance from MAFF in allowing NRA to influence strategic planning and land use issues. NRA officials had to be careful not to overstep the bounds of their acknowledged expertise as pollution not agricultural specialists (which was the role of MAFF's advisory arm, then the Agricultural Development and Advisory Service – ADAS – from 1997 the Farming and Rural Conservation Agency - FRCA). What they attempted to do, therefore, was to draw the environmental constraints of farm effluent management and disposal more centrally into production decision-making through the concept of Farm Waste Management Plans (FWMPs). The idea was a simple one of identifying, on a map of the farm, the areas suitable for waste disposal, taking account of topography, geology and soil type, the location of drains and watercourses, the season and rainfall. The chairman of the NRA's Farm Waste Group called for all farms to have a waste management plan, approved by the NRA, stating that in those cases where such plans indicated insufficient land capacity to absorb safely the waste produced, farmers would have to reduce their herds (ENDS, 1990). The plans, therefore, implied environmental limits to intensification on individual farms. Moreover, although focused at the farm level, if each farmer had to produce such a plan, it would allow the NRA to assess risks to the environment from livestock effluents at the catchment level;

the Authority's preferred level of operation. The consequence of such a policy would thus have been to draw this aspect of farm management into the NRA's catchment planning approach and to place clear environmental limits on livestock production policy. Rather than further promoting the technical fix mentality, embodied in the pollution control grants and the Farm Waste Regulations, FWMPs encouraged a more didactic, preventative approach. If accepted, the policy would have embodied an implicit acknowledgement of the legitimacy of the Authority to influence land use in order to protect water.

The Authority's ambitions with respect to agricultural policy were set out in detail in 1992 when it issued a major study on *The Influence of Agriculture on the Quality of Natural Waters in England and Wales*. While this welcomed government action relating to the new regulations, revised codes of good agricultural practice and grant aid for pollution control, it pressed for further initiatives, namely, "improved legislation, increased awareness by farmers of their responsibilities to the environment, and more effective measures to allow farmers to fulfil them" (NRA, 1992b, p. 11). The latter goal reflected an implicit critique of the agricultural policy framework under which farmers were then operating. In support of its criticisms the Authority referred to "increasing public and political concern" developing over "'some time' about the effects of modern farming methods on the environment" (NRA, 1992b, p. 11), highlighting in particular, a lack of consideration of environmental concerns in policies relating to livestock farming:

> "It is clear that when farmers increased their flock and herd sizes in
> response to government policies which promoted increased production,
> there was insufficient consideration of what to do with the huge amount of
> waste material so generated" (NRA, 1992b, p. 11).

Furthermore, the report claimed that the intensification of livestock farming had reached "a point where the associated land is often insufficient to cope with the slurry produced" (NRA, 1992b, p. 33). Anticipating its role in the new Nitrate Vulnerable Zones, the Authority proposed an extension of a land use control approach "to cover all 'sensitive' waters for a wide variety of contaminants" (NRA, 1992b, p. 12). Central to this concept was the key recommendation of the report, the launch of FWMPs, in which the Authority advocated liaison with MAFF to assess "the feasibility of using individual farm waste management plans" (NRA, 1992b, p. 82).

However, the Authority had come under strong pressures from agricultural interests, both behind the scenes and in the press, as this report was in preparation and its tone was surprisingly conciliatory. Even the FWMP proposal was somewhat deflated by MAFF's announcement, on the day of the report's publication, of the establishment of a pilot FWMP scheme (MAFF, 1992). Although what was proposed fell short of the NRA's ideal - in that plans were to be voluntary, not grant-aided and were not necessarily to involve expert input - it marked a certain accommodation of MAFF's and the NRA's positions and the Authority's acquiescence to MAFF's authority in this area.

The NRA and On-farm Regulatory Approaches

With respect to on-farm regulatory strategies, general NRA rhetoric suggested there might be a decisive move away from an 'information and persuasion' approach to agricultural pollution. Although a co-operative stance remained prominent, early declarations on the problem placed emphasis on the Authority as 'pollution watchdog'. In his foreword to the joint report by the NRA and MAFF on *Water Pollution from Farm Waste 1989*, the first produced by the NRA (in a dry year which witnessed a 30% decline in agricultural pollution incidents compared with 1988), Lord Crickhowell welcomed indications of increased environmental awareness and action in the agricultural community, pledging to continue to give "encouragement, assistance and advice to farmers". However, he also emphasised that such encouragement would be backed by tough legal action:

> "Prevention is much the best way forward but we have made it very clear to farmers and all others who may pollute our rivers that we shall not hesitate to prosecute offenders, and the courts have shown their willingness to impose severe penalties" (NRA and MAFF, 1990).

In the same report, the Authority stated that it was "determined to bring major or persistent offenders before the courts" and, while acknowledging the considerable variation in fine and cost levels associated with prosecutions across the country, dwelt in considerable length on just such a case taken in the Crown Court. This involved a major pollution incident, in which three million gallons of pig slurry polluted 60 km of river and killed more than 10,500 fish, and a persistent offender, already prosecuted four times in the magistrate courts. The farmer was found guilty, fined £10,000 and ordered to pay £20,000 in costs (NRA and MAFF, 1990, p. 5).

However, in later policy statements issued by the Authority, little was made of the 'watchdog' status of the organisation and prosecution of farm pollution was mentioned only in almost apologetic tones. Responding to the reduced numbers of water pollution incidents in 1992, Dr Petreath, Chief Scientist and Director of Water Quality, praised farmers in particular:

> "Great credit for this trend must go to the farming community: indeed major agricultural incidents continue to decrease - particularly those caused by storing silage." (NRA, 1993, Preface)

The "approaches required for preventing pollution and dealing with agricultural waste", the Authority declared, were "much broader than legal action" (NRA, 1993, p. 47). The overall tone of the section reporting on agricultural prosecutions in 1992 was much more conciliatory than previously, with less prominence given to legal action and examples cited "where formal action could not be avoided" (NRA, 1993, p. 47).

Table 4.1. Prosecution of farm pollution incidents in England and Wales and the South West#

	Under RWAs			Under NRA					
	1985	1986	1987	1988	1989	1990	1991	1992	1993
No of prosecutions instigated	159	128	225	148	161	123	159	c120	96
% of reported incidents	4.5	3.7	5.8	3.6	5.6	3.9	5.4*	c4.2	3.3*
% of serious incidents	28	21	23	16	31	19	N/k	N/k	N/k
% of major incidents	N/k	N/k	N/k	N/k	N/k	51	161	c179	152
	Under SW Water Authority			Under NRA South West					
	1985	1986	1987	1988	1989	1990	1991		
No of Prosecutions instigated	23	31	19	22	30	40	26		
% of reported incidents	3.7	3.7	3.4	3.2	5.4	5.1	3.3		
% of serious incidents	30	42	5	6	20	23	N/k		
% of major incidents					56	104			

Notes: # Rates relate to incidents occurring in the stated year for which prosecution proceedings have been started (whether complete or not). Prosecutions, in the stated year, of incidents occurring in a previous year are not Included. The 'rates' given for serious and major incidents in this table are also slightly misleading as they relate to the proportion of those categories which would be covered by the number of prosecutions taken. Thus, there is no certainty that, for example, 51% of major incidents were prosecuted nationally in 1990, only that the number of prosecutions would account for 51% of major incidents. More information is available for 1993 when only 21% of major incidents were actually prosecuted; the remaining prosecutions were for significant incidents. The figures, thus, have a tendency to overestimate the prosecution rates of serious and major incidents.
          * Rates given are of substantiated incidents.
Source: WAA and MAFF, 1987, 1988, 1989; NRA and MAFF, 1990; NRA, 1992c, 1992d, 1993, 1994a, 1995a; Environment Agency, 1996b; 1991 prosecution figures for South West Region supplied by Pollution Control NRA South West.

It is thus unsurprising that the **numbers** of farm pollution incidents prosecuted did not increase under the NRA; instead there was a decline to levels generally below those pursued by the RWAs in the mid-1980s (see Table 4.1). Furthermore, by encouraging the public to report pollution incidents and under increased vigilance from environmental activists, the Authority was faced with rising levels of pollution incidents not matched with a significant increase in the number of prosecutions (see Tables 4.1 and 4.2). This did not impress environmental groups and the NRA began to attract the type of criticism – of organisational laxness or even complicity – that had been directed at the RWAs. This culminated in the 1991 Commons Public Accounts Committee pronouncement that too few polluters were being taken to court and that farmers in particular were not being submitted to the 'polluter-pays-principle' (House of Commons Committee of Public Accounts, 1991).

The low numbers of prosecutions were partly accounted for by a fall in the number of the most serious farm incidents from 1991 (see Table 4.2). Indeed, between 1991 and 1995 numbers of 'major' agricultural pollution incidents fell from 99 to 28, with a corresponding fall in the contribution of agricultural pollution to all major water pollution incidents (26% in 1991 falling to 18% in 1995) (NRA, 1992d; Environment Agency, 1996b). However, an analysis of prosecution rates raises further questions about the NRA's commitment to a more legally oriented policy. The rates themselves are difficult to calculate because of changes in the categories distinguishing the severity of pollution and the move to identifying substantiated rather than reported incidents, introduced in 1990. At face value while there was no clear improvement in prosecution rates from 1989 to 1990 when the 'serious' category was in use, rates appeared to have improved substantially after the introduction of the new 'major' category (see Table 4.1). This change was accompanied by the introduction of national guidelines which advised that all major incidents be prosecuted, the only proviso being that there was "sufficient evidence to take such action" (NRA, 1993, p. 40). However, as the Authority itself acknowledged, the new major category represented "only the most severe incidents" and was not equivalent to the former 'serious' category (NRA, 1992c, p. 13). Yet by grouping the most severe incidents in this way and not distinguishing (except in 1993) which pollution categories prosecutions related to, the Authority was able to present the image of a tough regulator which prosecutes all severe pollution while not substantially increasing the number of prosecutions instigated. It could thus respond to the imperatives of the environmental lobby and counter its criticisms of organisational ineptitude while at the same time not antagonising the agricultural lobby. Towards the end of its life, any image of firmness towards agriculture was undermined as figures for 1993 were released in a form distinguishing the categories to which prosecutions related. These revealed that only 20.6% of the major farm pollution incidents which occurred in that year were subsequently prosecuted (NRA, 1994a). A later report by the National Audit Office stated that the NRA in 1993 prosecuted just 9% of the most serious farm pollution incidents (National Audit Office, 1995).

However, to reach the stage of legal action is an admission of the failure of other measures and the success of the NRA's protection of the environment might be better judged by reference to the level of pollution. The NRA adopted this view, claiming that improvements in river quality and a substantial reduction in major incidents indicated that its policies were proving successful (NRA, 1994b; 1995b). For example, in the River Torridge catchment, an early study of which helped construct farm pollution as an 'issue' (see Lowe *et al.*, 1997, pp. 75-77), the NRA reported a net improvement in water quality of 39% between 1985 and 1995, a reduction in farm pollution incidents from a peak of 93 in 1991 to 39 in 1996 and a 300% increase in the salmon rod catch between

1990 and 1995 (NRA, 1996, cited in Environment Agency, 1998, p. 156). The policy measures highlighted in this claimed success were those involving negotiation and persuasion, such as visits to farms and industrial premises, not court action. The continued pursuit of such persuasive strategies are examined below using case study material collected in 1990-1991 in the South West region, where the Authority faced major pollution problems caused by dairy farming.

Agricultural Regulation on the Ground in the South West Region

Under the RWAs the South West region had been a leading force in the development of regulatory strategies towards farm pollution, such as Farm Campaigns, with an emphasis on informal approaches and a lower than average prosecution rate. However, in the early years of the NRA, the region presented a tougher regulatory image, with a spokeswoman declaring, in the local newspaper, "We now have a higher rate of prosecutions than ever before and we are making sure the polluter pays . . . we are really cracking down now on farmers who pollute rivers." (*Western Morning News* 1 June 1990, p. 5). Early in 1992, the region also announced the deployment of special Pollution Task Forces, focused on the detection of pollution with a view to prosecution, which would 'blitz' known problem catchments (NRA South West, 1992). While Farm Campaigns continued, they were no longer the cornerstone of agricultural regulation in the South West. Instead visits related to anti-pollution grant approvals became a more important means of contacting and regulating farmers. Pollution Inspectors visited all farms where a pollution control grant had been applied for and subsequently revisited them to ensure that any approved work had been properly completed. Furthermore, a significant and growing amount of Pollution Inspectors' time was spent responding to pollution incidents.

However, the Pollution Inspectors who put the NRA's regulatory aspirations into practice operated not only within a framework of NRA policies and procedures but also through their own networks of colleagues, informants, activists and farmers. They thus had to negotiate conflicting discourses about the rights and wrongs of pollution. By the early 1990s Pollution Inspectors had generally been affected by the much-publicised moral mandate of the new Authority and increasing public concern over the environment. Rather than regarding pollution as merely an infringement of the rules, Pollution Inspectors increasingly saw it as morally wrong, an environmental 'crime'. Certainly there had been a fall in their threshold of tolerance of pollution from what it had been in the 1970s (Hawkins, 1984; Knowland, 1993). While acutely aware that effective pollution control involved much more than instigating legal procedures, Pollution Inspectors nonetheless felt there was a moral justification for increasing the number of prosecutions and recognised the benefits of such a strategy in terms of public support. One Pollution Inspector felt that whereas the public had had 'very little faith' in the former water authority, the NRA was more popular because of some 'crucial' prosecutions and the winning over of the press. Pollution Inspectors' views of pollution were framed by absolutes:

pollution was a 'dirty' river or stream and 'seeing a once dirty stream made clean' was their ultimate goal. In principal nothing less could be condoned.

Table 4.2. Farm pollution incidents in England and Wales and the South West

	Under RWAs			Under NRA						
	1986	1987	1988	1989	1990	1991	1992	1993	1994	1995
Reported	3427	3870	4141	2889	3147	N/k	N/k	N/k	N/k	N/k
Substantiated	N/k	N/k	N/k	N/k	N/k	2954	2770	2883	3329	2720
Serious	622	990	940	522	644	N/k	N/k	N/k	N/k	N/k
Major	N/k	N/k	N/k	N/k	239	99	67	63	36	32

	Under SW Water Authority			Under NRA South West			
	1986	1987	1988	1989	1990	1991	1992
Reported	830	666	836	589	782	N/k	N/k
Substantiated	N/k	N/k	N/k	N/k	N/k	718	686
Serious	74	422	420	160	173	N/k	N/k
Major	N/k	N/k	N/k	N/k	71	25	13

Source: WAA and MAFF, 1987, 1988, 1989; NRA and MAFF, 1990; NRA, 1992c, 1992d, 1993, 1994a, 1995a; Environment Agency, 1996b; 1991 prosecution figures for South West Region supplied by Pollution Control NRA South West.

In practice, however, Pollution Inspectors had and needed more workable definitions, shaped by problems they regularly faced in the field and with administrative and legal processes. In the South West the trend of less reliance than average on prosecution continued and while there was a small rise in the prosecution rates of serious then major incidents in the early years of the Authority, this increase lagged behind the national rate (see Table 4.1). On the one hand, the public stance of the NRA as an environmental guardian and increasing pressure from the rural public encouraged Inspectors to take an absolute view of pollution as wrong-doing, deserving of punishment. On the other hand, ambivalence over the deterrent value of prosecution related to still generally low levels of fines when compared with pollution prevention costs, inexperience in gathering evidence for prosecutions, an appreciation of the problems faced by farmers and the perceived need not to alienate them, encouraged more flexibility. In court magistrates seemed swayed more by arguments focusing on the farmer's (lack of) culpability than by the consequences of the pollution itself. Thus the moral discourse of courts focused on the farmer and not the pollution and fines were usually low. While average fines for agricultural pollution in the region rose from £136 in 1985 to £665 in 1991, these were well below the maximum that could be awarded of £20,000. It is unclear how much of a pollution deterrent such fine levels presented to a farmer who was being asked to spend thousands of pounds on effluent storage equipment to prevent pollution (WAA and MAFF, 1986; 1991 figures supplied by NRA South West Pollution Control Unit).

It is thus unsurprising that in the South West the NRA came under attack from local environmental groups over its operations. Friends of the Earth had set up a six-month campaign there in 1989-90 to publicise the farm pollution problem, bring pressure to bear on the NRA to be tougher in prosecuting offences and to draw up a manual to assist local groups elsewhere to combat such pollution. The organisation's view on farm pollution was clearly captured by its widely circulated poster featuring a dead fish and the caption, "FARM POLLUTION IS KILLING RIVERS!" - the explicit animism of which emotively encapsulated its view of the new environmental morality. FoE passed over the social origins of farm pollution and sought to operationalise an environmental morality in which all instances of pollution were prosecuted where the evidence justified bringing a case, whatever the particular circumstances (House of Commons Environment Committee, 1987, p. 162). The two campaigners employed by FoE monitored and reported the response time and overall performance of Pollution Inspectors with respect to pollution incidents, pushing Inspectors to undertake the formal water sampling procedures necessary for legal action. While reluctant to criticise the NRA in public, the FoE campaigners were dissatisfied with NRA strategy, believing that too much emphasis was still being placed on the "together we can beat it" approach which was seen as "too much carrot and not enough stick" (Interview with Devon FoE campaigner). More publicly FoE supported the Authority by calling for more resources and staff (Lowe *et al.*, 1997, pp. 154-161).

While Pollution Inspectors shared an environmentalist perspective on the farm pollution problem, they did not accept the single-minded approach to prosecution demanded by FoE. In fact, the FoE farm pollution campaign caused some resentment and irritation in NRA circles. One Pollution Inspector commented that, if FoE found something, they expected action on it immediately. He believed that they "don't really understand what's involved" in regulating farm pollution and instead simplified the issue.

It was not that Pollution Inspectors did not regard pollution as wrong but they realised that without the co-operation of the large majority of farmers, their task would be impossible. While a Pollution Inspector cannot be on each and every farm 24 hours a day, farmers are on the spot, actually managing potentially polluting practices, and their action or inertia can make a crucial difference between either a clean or a polluted river. In order not to alienate farmers, Pollution Inspectors realised they had to act in a way that the farming community regarded as reasonable and so had to negotiate between an environmental morality promoted by environmental interests and the 'moral economy' of farmers (see Lowe *et al.*, 1997, pp. 191-208). Farmers constructed moral discourses about the rural environment, albeit crucially different ones to environmentalists. Farmers' rural discourses were constructed around ideas of responsible production underpinned by a productivist morality, in which productive farmers were seen as better farmers and greater yields regarded as a sign of virtuous hard work, but qualified by a duty of stewardship. The technological changes of modern dairy farming were frequently associated with ideas of 'progress' and for the vast majority of these farmers the existence of pollution had not led them to question their agricultural practices. While most accepted the desirability of avoiding the pollution of rivers by farm effluents, the majority of farmers continued to regard pollution as the breaking of rules imposed by outside interests rather than an environmental crime, which they were morally bound to prevent. While deliberate pollution was condemned, accidental effluent spillage was generally regarded as an unfortunate by-product of modern dairy farming (Ward and Lowe, 1994). Not having embraced popular environmental morality but faced with a strong public indictment of farm pollution and polluting farmers, which challenged their moral economy, farmers' sense of their own worth and role in the countryside had begun to be undermined (Seymour *et al.*, 1997). In order not to alienate farmers further, and to secure their co-operation, the weapons of environmental morality that came to the fore in actions by Pollution Inspectors were generally not the most obvious ones of legal action. Instead Pollution Inspectors adopted the role of educators; a strategy equally imbued with environmental morality. On the one hand this helped reassure farmers of the value of their role and on the other attempted to mobilise a new sense of environmental responsibility and morality amongst them. The only farmers Pollution Inspectors sought to automatically prosecute in the courts were those responsible for severe pollution incidents. Beyond that, any farmers who were prosecuted tended to be those viewed essentially as 'rogue' farmers by both Pollution Inspectors and the farming community. Thus, on the majority of occasions, NRA Pollution Inspectors operated within the farming community's moral economy, working with the grain of what that community considered fair and unfair, right and wrong. Within this approach, while punishing farmers who deliberately emptied slurry into rivers was regarded as acceptable, prosecuting farmers whose equipment had overflowed or failed in heavy rain, who were actively trying to improve their effluent systems or who might be forced out of business by enforcement action, was much more morally questionable and not usually pursued (Lowe *et al.*, 1997, pp. 202-207).

Conclusions: a New Moral Agenda?

At its instigation the NRA was active in claiming to be a new moral force in environmental protection. It drew on the rhetoric of environmentalists and courted public support. It also set out to adopt a more environmentally oriented and publicly accountable approach than its predecessors. Legislation strengthened the powers of the new Authority, lending it the capacity to argue for higher fines and, with respect to agriculture, to press for precautionary action under the Farm Waste Regulations.

Yet while the NRA sought to identify with the environmental movement and draw on its moral authority, it was subject to other pressures and constraints which meant that its approach in practice was not so single minded or straightforward, particularly in relation to farm pollution. There was no metamorphosis of either attitudes or regulatory strategies within the organisation and it soon became evident that there was greater continuity than change in the way it dealt with industry in general and agriculture in particular.

While there was a greater willingness to take formal action in response to the most serious pollution, a shift to a strongly legally-oriented regulatory policy was resisted and considerable reliance continued to be placed on the types of information campaigns and persuasion that had been pursued by the former water authorities. In part this was due to its inherited legacy of much of the infrastructure, policy and staff of the former RWAs. Changing these structures and attitudes was a slow process, resisted by sections of staff at all levels, and continues to be a problem in the Environment Agency (Lowe *et al.,* 1997, pp. 191-208). More fundamentally, it reflected a concern within the NRA that too rigorous a policy of court action might alienate farmers and discourage preventative action at a time when anti-pollution grants were available and levels of the most severe agricultural pollution were beginning to fall. Educating farmers to be more effective environmental managers was regarded as a more positive approach although the Authority was frustrated in this goal by its lack of influence over production policy. While its new regulatory powers acted principally as a backup to the NRA's persuasive regulatory strategies, they were important both in terms of leverage they gave in dealings with polluters and in silencing criticisms from environmental groups.

However, despite general support from the public and environmental groups and new legislative initiatives, there was considerable outside resistance from other quarters to the NRA's moral imperatives. Although provided with an entrée into agricultural policy circles by the 1989 Water Act provisions, NRA influence over production and land use policy was circumscribed. Productivist agricultural interests continued to dominate the agricultural policy community and the NRA. However, the environmental morality, which the NRA promoted, never became a central force. Furthermore, farmers on the ground were only slowly coming to see pollution as an environmental crime rather than a breaking of the rules and to develop the environmental awareness necessary to manage their systems safely. In court, despite the tenfold increase in the maximum fine that could be imposed for water pollution to £20,000, the legal system failed to take environmental morality on board and to impose high fines for pollution. This powerful outside resistance to the NRA's moral imperative is more easily accounted for if we consider the history of the organisation's creation. The debate over water privatisation in the 1980s proved contentious, and when it was proposed that regulation should be placed in hands of private companies, established policy communities were thrown apart and wide-ranging support formed for an independent regulatory authority. The outcome was the creation of the NRA which must surely be seen as one of the greatest anomalies in the era of Thatcherite Conservatism (Maloney and Richardson, 1994). Since that time policy communities have regrouped and the NRA had to defend itself against being disbanded when the initial proposals for an integrated Environment Agency were set out. In such a hostile environment, it is unsurprising that the NRA's approach was one of reform and education rather than revolution.

Although the NRA ultimately failed to establish environmental morality as an underlying value of rural policy, it did introduce new environmental values into water regulation and management and helped strengthen environmental policy initiatives. It also presented an important challenge to the productivist approaches dominant in MAFF. Since the creation of the Environment Agency, concerns about agriculture have been submerged within the new organisation under a heavy workload related to waste management. Despite fears that a legacy of dramatic effluent store failure may have been built under the anti-pollution grant aid policy of the early 1990s and worries over diffuse agricultural pollution (Ward *et al.,* 1998) as farm pollution incident figures have improved, agricultural work is not likely to assume its former prominence of the late 1980s and early 1990s. Pressure for change at MAFF though is building. Weakened by the crises of environmental degradation, food surpluses and health scares, such as BSE, and more recently international agendas promoting sustainable development, productivist approaches continue to be challenged. Only recently a UK Round Table on Sustainable Development report (1998) called for changes in the culture and structure of MAFF to reflect the changing agenda of rural and agricultural policy, particularly with respect to rural development and environmental concerns, in order to enable the Ministry to play a more effective role in furthering sustainable development. With the current preoccupations of the Environment Agency, any far-reaching integration into rural initiatives of the environmental morality championed by the NRA is most likely to occur under the auspices of international initiatives linked to sustainable development and Agenda 21.

Chapter 5

Farmer Adjustments to Tobacco Policy in Southwest Virginia

John T. Morgan

Introduction

Government policy exerts considerable influence on the patterns of agricultural production in the United States, and the production of tobacco is strictly controlled by federal legislation. Franchises or allotments to produce a certain area or weight of tobacco are granted to farms based on history of production of the crop. Modifications to federal tobacco programmes have occurred from time to time, and farmers have often reorganised their operations to adjust to programme changes. Burley tobacco production controls have traditionally restricted allotments to particular farms. Since 1971, however, numerous burley tobacco production units have been consolidated under the 'lease and transfer' provision of the current federal tobacco programme. This paper traces the development of the burley tobacco programme and examines adjustments by Southwest Virginia farmers to the lease and transfer provision. The study focuses on Washington, Scott, Russell, Lee and Smyth counties, which produce almost all of the state's burley tobacco (Fig. 5.1). Extensive fieldwork was conducted in Washington County, Virginia's leading burley producer.

Southwest Virginia is part of the Appalachian Highlands, and almost all of the region's tobacco is grown in the Valley and Ridge physiographic province. This province, which traverses the area in a North-East to South-West direction, is bounded on the South by the Blue Ridge Mountains and on the North by the Cumberland Plateau. The broadest valley, the Valley of Southwest Virginia, is underlain by limestone rock while in some of the region's narrow valleys the subsurface rock is shale interbedded with limestone.

The great majority of the region's residents live in rural areas or in small urban settlements. The City of Bristol, and Washington and Scott counties, however, form part of the Tri-Cities (Johnson City - Bristol - Kingsport)

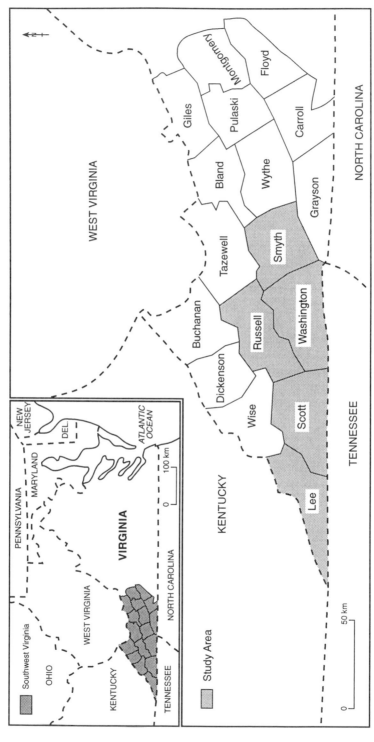

Fig. 5.1. The Counties of Southwest Virginia

Metropolitan Statistical Area (MSA), which has a population of about 500,000, most of whom live in the state of Tennessee. Most of the rural members of the area's workforce are non-farm residents who commute to work in the region's cities and small towns. In addition to workplaces in the Tri-Cities MSA, the towns of Abingdon, Lebanon, Marion and Gate City offer residents a variety of employment opportunities.

During the last few decades, the spread of non-farm residential development, including a proliferation of rural subdivisions, has been significant in the study area. At least two of the counties, Washington and Smyth, form a part of the recently designated 'Spersopolis,' a region of dense rural housing that includes much of the Piedmont of the Carolinas and Georgia, and the Valley of East Tennessee and Southwest Virginia. Associated with the rural non-farm population of Spersopolis are high levels of long-distance commuting, a dense concentration of manufacturing workers and a great density of dispersed mobile homes (Hart and Morgan, 1995).

Agriculture maintained a predominant position in the economy of rural Southwest Virginia until the middle of the twentieth century, but has declined greatly since World War II. The area's economy has shifted from self-sufficient agriculture to a mixed economy in which manufacturing is dominant and part-time farming plays a key role. At present the agriculture of the region is characterised by small farms and the important enterprises are dairying, livestock farming and burley tobacco, which is the principal cash crop of the study area counties. The future of tobacco farming in the region is in jeopardy, however, as health, social, political and economic forces pose an imminent threat to an activity that has become a 'way of life' for a multitude of small farmers (Wilson, 1989; Mundy and Purcell, 1996; Evans, 1997; Tennis, 1997).

Burley Tobacco

The United States Department of Agriculture (USDA) designates six major classes of tobacco based on variations in soils, climate, varieties of seed, cultural practices and curing methods. Air-cured tobacco is designated Class 3 by USDA and within Class 3, burley tobacco is classified as Type 31. In addition to Southwest Virginia, burley tobacco is produced in Western North Carolina, East and Middle Tennessee, and in several regions of Kentucky, the leading burley producer (Doub and Crabtree, 1973; Algeo, 1997).

Few crops are more labour-intensive than tobacco, and burley tobacco remains even more dependent on human labour than flue-cured tobacco, which has seen several of its processes mechanised over the past couple of decades (Hart and Chestang, 1978, 1996; Morgan, 1978). A brief discussion of the most important traditional processes required to harvest a crop of burley aids in understanding the nature of tobacco farming in Southwest Virginia.

Burley tobacco that matures to be harvested is transplanted from seedbeds that are prepared in February or March. The young plants require three months of growth before they can be 'pulled' from the bed and 'set' in the field. During the last five years, Southwest Virginia farmers have begun to rely more on greenhouses with waterbeds to produce plants than the traditional outdoor seedbeds. Although greenhouse plants are more expensive, they are stronger than plants from traditional beds, and they can be pulled and transported much more easily than those grown outdoors. Most Southwest Virginia farmers do not have greenhouses on their farms, but purchase waterbed plants from either large farmers who produce more plants than needed for their farms, or from greenhouse growers (within or outside the region), who grow plants primarily for commercial sale (Morgan, 1996).

Transplanting of tobacco usually takes place between 20 May and 20 June. Before setting takes place the land is broken, disked and harrowed, and treated with chemicals to prevent weed and pest growth and to combat certain diseases to which tobacco is susceptible. Heavy applications of fertiliser, typically 2.2 t/ha, are also applied before setting. Most farmers now rely on tractor-mounted or pull-type mechanical setters, rather than hand transplanters, to assist labourers in the task of transplanting. About 20,000 plants are needed to set a hectare of tobacco with the plants spaced 5.0-5.5 cm apart in rows separated by a width of 105 cm.

A crop of tobacco is usually ploughed 2-4 times before it is too tall for a tractor to pass over the tobacco plants. After 60 or 70 days of growth, the flowers of the tobacco plants are removed in a process known as 'topping,' which forces plants to grow more horizontally than vertically. Plants are usually topped at a height of 20-22 leaves. After topping is completed, a chemical growth inhibitor (or 'sucker control') is applied to prevent the growth of suckers at the intersection of the tobacco stalk and the leaf stem. Farmers usually apply sucker control with a compressed-air hand sprayer mounted on their backs. Large-scale farmers are beginning to use specialised 'hi-cycle' or above-crop sprayers. The same methods are used to spray the crop for two of types of worms that can do great damage to a crop in a short period of time.

Once the crop is mature, usually in early September, burley tobacco is harvested by cutting the individual stalks near the ground and 'spearing' them onto 1.4 m wooden sticks. After a 5-7 day period of moisture loss in the field, the tobacco is hung in an air-curing barn where it remains until the curing process is complete, usually by the middle of November. Once curing is completed, tobacco is taken out of the barn, stripped from the stalk, sorted or graded into three or more quality groupings, pressed into 50 kg bales and taken to market. The tobacco market sales begin about 15 November and each year's entire crop is usually sold by the middle of January.

The Burley Tobacco Programme

Legislation to control agricultural production in the United States in the 1930s was a response to overproduction and low prices. Burley tobacco production has been subjected to government restrictions since 1934, when the Agricultural Adjustment Administration initiated crop control programmes. Tobacco controls from 1934 to 1971 were based on area of production, and tobacco allotments were associated with the farm, rather than the farmer. Base areas were set by the United States government for each farm by reference to 1931-1933 production. The government made cash payments to farmers participating in the programme. During the 1940s price supports were added for tobacco. If the bid price on a lot of tobacco is less than the government price-support level, the farmer can accept the government rate (Durand and Bird, 1950; Doub and Crabtree, 1973).

Because of wartime demands for burley tobacco the area quota was increased by 10% in 1943 and 20% in 1944. Moreover, provisions were added that (1) made the minimum allotment per farm 0.4 ha (1 acre), and (2) relaxed regulations regarding the creation of new farms. Many new minimum-size burley allotments of 0.2 ha or 25% of the cropland were created during the war years. As a result of a decline in post-war demand for burley tobacco and increased yields, allotment cuts were made for the 1946-1947 marketing year. The minimum 0.4 ha provision was not deleted, however, and allotment cuts were therefore effected primarily on large allotment holders (Mann, 1975).

Rising yields brought about additional quota reductions, all of which were imposed on medium- and large-scale operators. From 1945 to 1951 allotments greater than 0.4 ha were cut by 32% whereas smaller allotments suffered virtually no reductions. In 1953 a 20% allotment cut was enforced, and the minimum allotment was reduced to 0.3 ha (seven-tenths of an acre). Many farmers, however, took advantage of a loophole in the programme to increase their area in spite of allotment reductions. Farmers were allowed to overplant their allotments, providing they paid a penalty on the excess tobacco produced. Overplanting was a strategy to enlarge their future base area (Mann, 1975).

So many producers were building up their allotments by overplanting that by 1955 it was a major threat to the programme. A bill was passed prohibiting future overplantings, but farmers were allowed to retain areas gained through past overplantings. The bill also reduced the minimum allotment to 0.2 ha. Subsequent reductions increased greatly the number of farmers holding minimum area allotments.

From 1955 to 1971, the only way a farmer could increase the area of tobacco was to buy a farm with allotment or to rent additional land and cultivate it on the farm with which it was associated. A farmer would probably have to rent several dispersed plots of tobacco to achieve the desired scale of production. By farming dispersed plots the farmer's production costs per hectare increased significantly. Such costs could be reduced greatly if the farmer were allowed to lease allotments and consolidate them on one or two farms.

In 1971, burley tobacco growers approved a new programme that replaced area controls with weight controls, and allowed within-county lease and transfer of allotments. And, to quieten fears of the holders of small allotments that holders of large ones would over-consolidate, a 6.8 t maximum lease provision was imposed. In 1984, the maximum lease was increased to 13.6 t, with the actual amount that an individual farm could add determined by the total amount of cropland on the farm. In 1991, a provision was added to the programme that allowed the permanent sale of weight quota from one farm to another. Allotment holders must grow their quota (actively or through lease arrangement) in two of three years or forfeit the franchise to grow burley tobacco (Morgan, 1996).

Impact of Lease and Transfer

If one examines the data in Table 5.1, it becomes obvious that the farmer needs to be able to consolidate tobacco weight quota to achieve an adequate scale of production. One-third of the farms in Washington County have quotas of no more than 0.5 t (less than 0.2 ha), and more than half the farms in the county have quotas totalling 5.9 t (about 0.2 ha typically) or less. Moreover, 85% of farms have quotas of no more than 1.1 t, which can be produced on about 0.4 ha of land.

Quota holders have participated strongly in leasing activities since the programme's inception. Many farmers want to increase their tobacco weight, and are willing to pay to lease quota; even more quota holders are willing to relinquish their franchises, especially if the lease rate is attractive. The gradual decrease in tobacco farm numbers in the region during the period 1969-1992 is an indicator of farm consolidation through lease and transfer (see Table 5.2). The decrease in farm numbers in the region during the period ranges from a 31% loss in Washington County to 46% in Scott County.

Table 5.1. Size groups of Washington County, Virginia tobacco farms (quota holders), 1995

		kg		Basic Quota	
No. Farms	%	From	To	kg	(%)
1,102	33.0	1	454	302,777	12.0
665	20.0	455	589	346,226	14.0
1,054	32.0	590	1,134	827,302	33.0
393	12.0	1,135	2,269	599,370	24.0
85	2.6	2,270	4,539	255,884	10.0
14	0.4	4,540	11,349	90,659	3.6
4	0.1	11,350	34,049	70,592	2.8
3,317				2,492,810	

Source: United States Department of Agriculture, Agricultural Stabilisation and Conservation Service, Washington County, Virginia

The extent to which leasing is taking place is also indicated in Table 5.3, which shows that in 1994 only 35% of Scott county quota holders actually marketed tobacco, and less than half the quota holders in Washington and Smyth counties actively participated in tobacco farming by marketing the crop. Table 5.3 also indicates that most of those leasing-out production quantities have small quotas, because in Smyth County (579 leases; 307,666 kg leased-out) the average lease was for 531 kg (about 0.2 ha). The degree to which the typical tobacco farm expanded in size in 1994 might be shown by comparing with data for Scott County. The average quota in the county is 609 kg, but after leasing-in weight the average farm increases by more than three times to about 1,886 kg, or an increase from just 0.2 to 8.1 ha.

Table 5.2. Number of tobacco farms in Southwest Virginia 1969-1992

Year	Lee	Russell	County Scott	Smyth	Washington
1969	1,591	1,165	2,138	876	2,140
1974	1,275	1,034	1,393	614	1,821
1978	1,254	980	1,515	618	1,855
1982	1,379	1,038	1,663	691	1,936
1987	1,119	837	1,274	554	1,515
1992	974	792	1,151	510	1,484
% Change 1969-92	-39%	-32%	-46%	-42%	-31%

Source: United States Census of Agriculture (1969-1992)

Table 5.3. Burley tobacco programme statistics, three Southwest Virginia counties, 1994

	Scott	County Smyth	Washington
Farms with quota	3,522	1,273	3,563
Basic quota in kg	2,353,623	734,373	2,472,859
Leases approved	-	579	-
Weight Leased	-	307,666	-
No. farms marketing tobacco	1,248	583	1,538
% farms	35	46	43

Source: United States Department of Agriculture, Agricultural Stabilisation and Conservation Service, Scott, Smyth, and Washington counties, Virginia

Farmer responses to the lease and transfer provision have produced several significant changes on the landscape of Southwest Virginia. The consolidation of allotments or quota to a continually reducing number of farms has altered the image of the region. Travelling through Southwest Virginia during the early 1970s, small, often garden-sized tobacco patches were a ubiquitous feature in much of the region. Now the fields are fewer in number but much larger in size.

Lease and transfer has also created a new geography of tobacco production in the region. Marginal areas, such as mountain slopes and hillsides, have lost tobacco to the better soils of the broader valleys. Even in the broadest valley, the Valley of Southwest Virginia, there is not a uniform distribution of tobacco. One can drive down a particular road and find farmland but virtually no tobacco, whereas farmland along another road is littered with tobacco fields.

Farmer Responses

Burley tobacco farming is dominated by part-time farming, and lease and transfer has allowed part-time farmers to increase the size of their tobacco farms and at the same time to improve the efficiency of their operations. One such farmer is 41 year old Phil Henderson, who has been a part-time farmer since graduation from high school. Henderson commutes 40 km from his Meadowview, Virginia home to Bristol, Virginia where he is employed full-time as a maintenance mechanic at Bristol Compressors, a manufacturing plant. In addition to the four twelve-hour shifts (06.45 to 18.45 hrs) that Henderson works each week, he also does 'custom' landscaping with his Massey-Ferguson backhoe and Komatsu tractor with front loader. His wife Janie works during the school year in the cafeteria of the local elementary school.

Owning no farmland, Henderson leases a 17 ha farm, on which he grows 0.6 ha of tobacco and 4.1 ha of hay for his three riding horses. He subleases about 10.1 ha of the farm to a local dairy farmer who grows corn there. Henderson usually produces 1,180 kg of tobacco, 953 kg being the basic quota on the leased farm and 227 kg being leased-in from his father's farm. Henderson has farmed under this arrangement since 1987. For more than a decade before that time, he leased his father's quota only. Henderson engages in a co-operative labour exchange with a friend, who helps with planting, topping, harvesting, stripping and grading of the crop. The two friends are also co-owners of a tobacco setter and a backpack sprayer. Henderson hires two co-workers from Bristol Compressors to assist with cutting and spearing the tobacco. Tobacco has provided gross annual revenues of more than $4,000 for Henderson in each of the last three years.

Another part-time farmer is Freddie Rouse, who lives with his family on 1.4 ha in the village of Glade Spring, Virginia. Rouse commutes 3.2 km to work at Emory and Henry College where he serves as the Chief Security Officer. His wife Judy also works full-time at the college, as a custodian. Freddie Rouse works Monday through Friday from 06.00 to 14.00 hrs, a schedule that leaves him time to do farm work. Rouse, 43, has farmed tobacco for 17 years. His 1.4 ha in Glade Spring has a basic quota of 266 kg, and he leased-in 1,135 kg for the 1996 crop. All the tobacco is on one field of 0.6 ha. Rouse also leases his parents' 14.6 ha farm in Widener's Valley, 12.8 km away. Twelve hectares of the farm are in pasture, on which his twenty head of beef cattle graze; another 1.6 ha are planted to oats, which are ground into feed for the cattle. The farm has a tobacco quota of 227 kg, to which Rouse adds 454 kg of leased tobacco. All the tobacco on the farm is grown on a 0.2 ha plot. Rouse owns a new $15,000 tractor with equipment and has a mechanical tobacco setter. He engages in co-operative labour arrangements with his brothers-in-law, who also grow tobacco. Rouse received nearly $9,000 for the 1996 tobacco crop.

Although the great majority of Southwest Virginia farmers are part-time operators, each community has some full-time farmers. Dairy farms are almost always full-time operations, and there is a scattering of full-time farms dominated by burley tobacco. Dairy farms rarely focus on tobacco production, and if they do, it is likely that the farm will soon shift away from dairying and place greater emphasis on tobacco production. Full-time tobacco farms also usually place considerable emphasis on beef cattle grazing. A few of the farms have used the lease and transfer mechanism to amass enough production quota to become burley tobacco 'super farms'. Although relatively few in number, these farms are important because they control a significant portion of total tobacco production (perhaps 20%) produced each year, and their operators tend to be the innovators who try new and different farm strategies, often copied later by smaller farmers.

One tobacco 'super farm' is operated in the Cedarville community of Washington County by Jackie Thomas and Bill Neese, who operate their individual farms but are partners in the production of tobacco and tomatoes. Thomas, 54, closed down a dairy operation in 1990 and shifted primarily to tobacco production. He owns a 44.6 ha farm and leases four additional farms, totalling 202.5 ha. Much of the farm is devoted to pasture for his 100 head of beef cattle, but about 81.0 ha of hay are also harvested each year. Thomas has an equipment inventory of about $300,000. He employs one full-time worker on the farm. Bill Neese owns only 4.1 ha but leases five farms with a total of 182.3 ha. Neese devotes most of the farms to pasture for his 60 head of cattle, but also grows 60.8 ha of hay.

The main reason for Thomas and Neese leasing a total of nine farms is to have farms to which large amounts of tobacco can be legally attached through leasing. Each of the nine farms has enough cropland to qualify for the maximum lease allowable (13,620 kg). Jackie Thomas has purchased outright extra tobacco production quota during recent years to bring his basic quota up to 6,810 kg. Bill Neese has done the same and now has a quota of 3,632 kg. In 1996 the two men added more than 100 leases totalling 81.7 t of tobacco to their quotas. What had been more than 100 small tobacco farms during the pre-lease and transfer period were combined into one farming operation. The Thomas-Neese tobacco operation of 24.3 and 90.8 t typically grosses nearly $400,000 per year.

The key to operating a large tobacco farm is availability of labour. Farmers who rely on local labour can probably grow no more than 4.1 ha of tobacco. Those who grow more than 4.1 ha have come to rely on migrant labourers to harvest and process the crop for market. Mexican migrants have provided much of the labour for the large tobacco farms during the last decade. During late July and August, Thomas and Neese employ eight Mexican men to pick their 2.4 ha of tomatoes and, to a lesser extent, to work in tobacco fields in pre-harvest activities. When tobacco harvest time comes in late August, 16 more Mexicans are added to the Thomas-Neese labour force. The 24 labourers cut and spear tobacco during late August and throughout the month of September, and in early November they begin to strip, grade, and bale the cured tobacco for market. Thomas and Neese try to have the entire 90.8 t marketed before Christmas, but they sometimes do not finish selling until the middle of January. The old Thomas homeplace serves as the dormitory for most of the migrants during their stay on the farm.

Concluding Remarks

The Burley tobacco programme has been successful because its support price system guarantees the farmer a reasonable return on the investment required to produce such a labour-intensive crop. The lease-and-transfer provision of the Burley tobacco programme has allowed farmers to consolidate allotments to achieve almost any scale of production desired. The small part-time farmer, such as Freddie Rouse, has been able to enlarge his tobacco farm from a few hundred kg and revenue potential of less than $1,000 to a few thousand kg and a revenue potential of several thousand dollars. Others, including the large part-time farmer who wants to grow 2.0 ha of tobacco, the small full-time farmer who desires 4.1 ha of tobacco, and the super farmer who wants more than 20.3 ha, have all been able to lease enough tobacco to achieve the desired scale of operation. Those desiring to lease-in tobacco quota are able to do because the majority of quota holders opt to lease-out their production quota for a relatively small fee rather than face the rising costs and labour shortages associated with tobacco farming today.

Lease-and-transfer has produced significant landscape change in Southwest Virginia by concentrating tobacco in larger fields on fewer farms that are typically located in areas better suited to produce the crop. The shift of tobacco production to larger farm units may be a prelude to even greater changes on the horizon. Although the tobacco programme has been beneficial to the small farmer and has largely been financed by fees paid by the farmer, the programme's elimination by the US Congress appears to be imminent. The highly publicised health risks associated with the consumption of tobacco products make continued support of the programme unpalatable to legislators who do not represent tobacco growing areas. If the tobacco programme is eliminated, the small tobacco farmer in Southwest Virginia is likely to be eliminated as well, with the production of the crop concentrated on the farms of a few large operators. And, unfortunately, there are no viable agricultural alternatives to tobacco for the small farmer in Southwest Virginia.

Chapter 6

A New Space or Spatial Effacement? Alternative Futures for the Post-productivist Countryside[1]

Keith H. Halfacree

Introduction: 'The Land is Ours'

On 23 April 1995 around 400 people - a motley collection of anti-roads protesters, New Age travellers and frustrated smallholders - began a week long occupation of a disused airfield, 12.2 ha of set-aside land and a copse some 4.8 km south of St George's Hill, near Weybridge, Surrey. Over the following week they built a village of communal and individual huts and tents, and had set up a bread oven, bath house and sweat lodge. Most importantly, the group cultivated some of the set-aside land, planting broad beans, marrows, cauliflowers and a tree on St George's Hill itself (*Guardian,* 1995; *Independent on Sunday,* 1995). The site of this demonstration was not accidental, as the demonstrators saw themselves as heirs to the 17th century Diggers and it was on St George's Hill in 1649 that Gerrard Winstanley and the original Diggers launched their campaign for an end to land enclosure and access to the land for the ordinary people of England.

The occupation of the land in 1995 was part of a new campaign, launched as a result of a newspaper article earlier that year (Monbiot, 1995) and now officially endorsed by Oxfordshire County Council. As the key figure in the campaign, George Monbiot, argued, the occupation should not be seen as a protest against anything per se (Monbiot, 1995) but as a reassertion of "people's control over how they live" (*The Guardian*, 1995). This new campaign is known as 'The Land is Ours' and sees itself as a "land rights movement for Britain". The key aims of the movement are shown in Table 6.1 and comprise an attempt to secure broader access to land in Britain, whether for homes, for employment or for popular access in general[2].

Taking a cue from 'The Land is Ours', this chapter argues that campaigns such as the one at St George's Hill can be interpreted, in part, as attempts to come to terms with the crisis of productivism in the countryside. This crisis has brought about a need to search for new uses and directions for the countryside in Britain and elsewhere. However, the campaign is representative of just one way in which the contours of the post-productivist countryside may be shaped. Indeed, its chances of success are slim, as are the chances for any radical attempt to re-focus space in this way. This is because they are oppositional to a very solidly entrenched alternative organisational logic of space. Instead, new rural spaces are being induced from within this organisational logic, which may contradict and conflict with one another in many ways but which do not challenge this basic logic. Nonetheless, the crisis of productivism has allowed - created a conceptual space - for campaigns such as 'The Land is Ours'; whether they make any longer-term headway is a very political issue.

Table 6.1. Aims of 'The Land is Ours'

Land for homes: Use of ex-industrial sites for social housing, especially innovative and communal projects. Planning presumptions in the countryside to allow settlers to live on their own land and a new Caravan Sites Act to give travellers somewhere to live.

Land for livelihoods: Action to prevent the destruction of habitats and landscape features by intensive agriculture. Subsidies and planning to be redirected towards small-scale, high-employment, low-consumption land uses, such as organic smallholdings.

Land for life: Protection and reclamation of common spaces in town and country, and the end to enclosure of streets, playing fields, play and informal recreation areas, and city farms and allotments. A right of access to uncultivated land in the countryside. Reform of the planning process away from the developers' interests. A public registry of landownership and an introduction of community ground rents.

Source: 'The Land is Ours', 1995

The Post-productivist Countryside

The current crisis faced by agriculture and the suggestion that we may be moving from a productivist to some sort of 'post-productivist' regime in the countryside will be reasonably familiar to most of the readers of this chapter. The productivist era, for British agriculture at least, lasted from the end of the Second World War to around the late 1970s. Although under the shadow of the demands of the 'agricultural treadmill' (Ward, 1993), this era was characterised by a sense of security with respect to land rights, land use, finance, politics and ideology, all bolstered by a highly corporatist relationship between the agricultural industry and the British government. In short, agricultural predominance in the countryside was regarded as benign, with agriculture being promoted as a progressive and expanding food production orientated industry (Marsden *et al.*, 1993). Crucially, other institutions not directly related to agriculture but also concerned with the British countryside tended to accept this role for agriculture. For example, the 'agricultural exceptionalism' (Newby, 1987, p. 216) of the 1942 Scott Report, whereby agriculture was virtually exempted from planning controls compared with other forms of industry, sent a clear message that agriculture's hegemonic position within the countryside was both proper and there to stay.

By the 1970s, however, this position was breaking down. There is not the space here to detail the threats which the productivist regime faced, suffice to note Marsden *et al.*'s (1993) stress on the economic and other contradictions in the 'Atlanticist' food order, the post-*Silent Spring* concern with the negative environmental effects of many agricultural activities, and the impacts of service class in-migration to rural areas (Cloke and Goodwin, 1992, pp. 327-328). As a result of these contradictions, British farmers now both feel and indeed are in a much less secure and certain position as regards their role in both agriculture and rural life generally. Moreover, this is also the perception of the general public, who have increasingly questioned the role of farmers as guardians of the countryside, and the financial and other support that this entails. Thus, we see high levels of debt and depression in the farming community, sometimes culminating in suicide, and a growing involvement in non-food producing activities such as Bed and Breakfast accommodation.

The crisis in productivism is not just of significance to the agricultural population, as it also signals a crisis for all of those implicated in the productivist era. It suggests that the hegemonic domination of rural areas and rural society by agriculture, which of course extends further back than 1945, is itself coming to an end. Hence, post-productivism signals a search for a new way of understanding the countryside. New interests and actors are coming on the scene in an attempt to create a rurality in their image. Spurred on by a recognition of supposedly 'surplus' land, there has emerged: "a wider debate on the access of non-agricultural interests to rural land, thereby allowing a wider range of interests to stake claims ... further compromising the productivist ideology which is now so obviously in disarray ..." (Marsden *et al.*, 1993, p. 68). This debate is concerned, fundamentally, with the production of space.

The Production of Space

The Abstract Space of Capitalism

Lefebvre's *The Production of Space*, available in English in 1991 but first published in French as long ago as 1974, charts the historical development of space in order to stress its increasingly produced character[3]. By 'produced', Lefebvre is drawing attention to the way in which space is now intrinsically linked with capitalism and exchange, rather than being somehow relatively independent of society's mode of production. Specifically, under capitalism, space becomes 'abstract', highly flexible and contingent, moulded by the pressures and demands of the market and social reproduction. Consequently, space initially subsumes place, with the loss of essential meanings described so evocatively by Relph (1976) and others.

In spite of the homogenisation of capitalist space in terms of its reduction to abstract exchange value relations, this space is of course only 'conceptually emptiable' (Sack, 1986, p. 37). It is always material, filled 'on the ground', leading to the tensions between this fixity and its abstract fluidity. Thus we have 'place' re-emerging where the spatial flows of capitalism are 'stopped' and 'situated' (Merrifield, 1993, pp. 525, 522). Lefebvre notes the contradiction thus:

"Space ... is both abstract and concrete in character: abstract inasmuch as it has no existence save by virtue of the exchangeability of all its component parts, and concrete inasmuch as it is socially real and as such localised. This is a space, therefore, that is homogenous yet at the same time broken up into fragments"(Lefebvre, 1991, pp. 341-342).

Focusing on the material form which space takes, attention is drawn to how geographies of capitalism - economic geographies, social geographies, etc. - provide the means of economic and social reproduction. However, we can go further, showing how these material geographies combine with what may be termed 'imaginative geographies'. This requires a broad conception of space, such as that also provided by Lefebvre.

The Conceptual Triad of Space

Lefebvre (1991, pp. 33, 38-39) recognises three key dimensions to space, a 'conceptual triad' (see also Harvey, 1987, pp. 265-269; Merrifield, 1993, pp. 522-527):

Spatial practices: actions - flows, transfers, interactions - which 'secrete' that society's space, facilitating socio-economic reproduction. These practices are linked to everyday perceptions of space, and to the rules and norms which bind society together.

Representations of space: formal conceptions of space, as articulated by planners, scientists and academics in general. A conceived and abstract space but expressed directly in such things as monuments, the workplace and bureaucratic rules.

Representational spaces: the diverse and often incoherent images and symbols associated with space as it is directly lived. Space as symbolically appropriated by its users. A clandestine and potentially subversive space within typical everyday experience.

The critical tension in this conceptual triad is between the representations of space, which reflect power and the interests of the dominant groups within society, and the representational spaces, which are "the dominated - and hence passively experienced - space[s] which the imagination seeks to change and appropriate" (Lefebvre, 1991, p. 39). Alternatively, but not overlooking the power relations inherent in this division, these two types of space can be considered in terms of two types of discourse. Representations of space conform to 'academic discourses', the constructs of those attempting to understand, explain and manipulate the social world, carefully formulated and organised (Sayer, 1989). In contrast, representational spaces have more in common with Sayer's 'lay discourses', the less formal, relatively unexamined constructs used within everyday life.

Place, Structured Coherence and Productivism

Whilst Lefebvre distinguishes the triadic nature of space in this way he is keen to promote an integrated analysis. This is because the three dimensions of space are brought together in place: "the 'moment' when the conceived, the perceived and the lived attain a certain 'structured coherence'" (Merrifield, 1993, p. 525). The term 'structured coherence' is taken from Harvey (1985, pp. 139 ff.) and, drawing upon regulation theory, refers to the way in which the economy, State and civil society cohere in a **relatively** stable fashion at the local level[4]. In other words, spatial practices, underpinned by representations of space and engendering representational spaces, tend to sustain themselves in a relatively stable manner over time.

Productivism can thus be seen as being a key form of 'structured coherence' within rural Britain. Cloke and Goodwin have explicitly noted this: "Most rural areas have exhibited **localised** structured coherences in recent history, based on the central position of agriculture ... in local society, economy and politics" (Cloke and Goodwin, 1992, p. 327). With post-productivism there is taking place a transition to a new form of structured coherence within the rural sphere. The breakdown of productivism represents a failure on the part of the spatial practices and the representations of space to explain and legitimate themselves and the emergence of oppositional representational spaces that this failure encourages. For Cloke and Goodwin (1992, pp. 328-333), the key forces bringing about the, as yet uncertain, new coherences which will reassert authority over rurality are those of:

Economic restructuring: notably the increased commodification of space;

Socio-cultural recomposition: especially migration linked to the appeal of the 'rural idyll';

The State's position in managing and directing this change.

Three types of potential structured coherences are making their presence felt. The first two involve maintaining the physical rural space that has been carved out by the productivist era, whilst the third sees a downgrading and transgression of the rural-urban distinction as conventionally understood.

Alternative Futures for the Countryside

New Spaces for Old?

The first two alternative futures for the countryside are linked to post-1945 productivism through the Scott Report of 1942. In writing the report there was a divergence of views, between the majority, who favoured a bucolic future for agriculture refracted through a version of the 'rural idyll', and a minority, represented by Professor Dennison, who saw a more explicitly capitalist future for the industry (Newby, 1979, pp. 230-231). The latter view anticipated the 'super-productivism' of agribusiness.

Super-productivism

With productivism, the key spatial practice concerned agricultural food production, with the attendant representation of space being that of the countryside as a food production resource. For a re-statement of this unadulterated productivism in the (hence, misleadingly titled) post-productivist era - a 'super-productivism' - we only have to look at the activities of agribusiness. Here, the representation of space and everyday representational spaces have been shorn of any 'moral' stewardship element and the land is treated solely as a productive resource linked to profit maximisation. Indeed, such is the physical impact of agribusiness that the representational spaces for those involved have little scope to diverge from the representations of space which guide and legitimate the industry, the capitalist rationality of agribusiness impinging in all directions.

The Rural Idyll

Agribusiness has met with considerable disquiet from rural and urban residents alike. In particular, this is because these critics uphold a very different future for the countryside. For them the predominant spatial practices are residence and attendant migration (counterurbanisation). The representation of space underlying these practices tends to be the 'rural idyll'. The characteristics of this idyll are well known (see Short, 1991). Physically, it involves small villages nestling in countryside that is farmed - by small non-capitalist-inclined farmers rather than agribusiness - and socially it involves 'community', friendship and belonging. The extent to which the rural residents' representational spaces conform to this spatial imagination vary (see Halfacree, 1995). However, as Murdoch and Marsden (1994) have shown, the new residents' spatial experiences are likely to be moving closer to the representation as the countryside is being re-made in this image (Halfacree, 1993, p. 34).

At first sight, the rural idyll and its attendant spatial behaviours and experiences may not appear to be as congruent with capitalism as the space of agribusiness. However, we can draw attention to a number of features which suggests a very happy marriage indeed. First, the rural idyll is a very 'conservative' vision, respectful of private property and 'traditional' institutions such as the nuclear family. It has little or no space for 'radical' challenges to the societal status quo, whether from socialists, feminists or radical groups more generally (see Halfacree, 1996). Secondly, a range of academics has associated counterurbanisation and rural living in the shadow of the rural idyll with middle class social reproduction (notably Thrift, 1987, 1989).

The importance of the rural idyll in leading the efforts to create a new structured coherence in rural Britain is clear (Cloke and Goodwin, 1992; Halfacree, 1995). For example, a key arena for the 'contested transition' (Marsden and Flynn, 1993) from productivism to post-productivism is the development process in rural areas. This has been demonstrated clearly in Murdoch and Marsden's (1994) wide-ranging study of Aylesbury Vale in Buckinghamshire, where they recognise that:

> "the crisis in agriculture has ... engendered much uncertainty about the proper uses to which rural space should be put. In this period of uncertainty conflicts around development processes have become increasingly bitter and have placed some amount of strain on the planning system" (Murdoch and Marsden, 1994, p. 29).

They go on to demonstrate the way in which contrasting images of the rural future have been materially produced: from the creation of new settlements, to extended suburbanisation, to golf course development. Overall, in spite of this variety, the Aylesbury Vale post-productivist countryside is seen to be increasingly moulded into a middle class space underlain by the rural idyll.

Although both congruent with capitalism, agribusiness and the rural idyll are not happy bedfellows, leading to much of the conflict over the environment which characterises the post-productivist rural experience. In many parts of Britain the rural idyll seems to have gained the upper hand. Thus, in Aylesbury Vale there is an "ascendance of certain aesthetic representations of the countryside over previous economic ones" (Murdoch and Marsden, 1994, pp. 215-216) and non-traditional farm buildings, the "stark monuments to the age of agricultural productivism" are frowned upon in the new countryside. Elsewhere, in places such as much of East Anglia, super-productivism may have a firmer and more accepted hold.

Alternative Futures for the Countryside

Radical Visions

Transgressing the Urban-rural Divide

In contrast to both super-productivism and the rural idyll are radical visions of new structural coherences for the countryside. These have emerged from tensions between the representations of space within productivism and the symbolic representational spaces. Here, the latter's subversive tendency is the key issue:

> "Differences endure or arise on the margins of the homogenised realm, either in the form of resistance or in the form of externalities ... What is different is, to begin with, what is **excluded** ..." (Lefebvre, 1991, p. 373).

Subversive representational spaces attack the key axis of domination expressed by spatial practices and representations of space, in their challenge to private property, exclusivity, hierarchy, etc. (Harvey, 1987, p. 266). Radical spaces are given an opening during periods of crisis, such as that currently being experienced by productivism.

Originating within such subversive representational spaces, the vision of the countryside proposed by 'The Land is Ours' and other radical rural movements conflicts not only with the other two rural futures but also with the spatial delineations of capitalism itself. Specifically, 'radical ruralities' involve a rejection of the sanctity of private property as represented in land ownership and advocate a move towards less capitalistic lifestyles and everyday priorities. Moreover, they do not just delineate the privatised spaces of the countryside but they also represent a transgression of the urban-rural division as it is conventionally understood. For them, what is suitable for the countryside is also suitable for the city; hence the geographical coverage of 'The Land is Ours' to encompass the urban as well as the rural environment.

For these radical visions of the countryside, the predominant spatial practices are likely to revolve around decentralised and relatively self-sufficient living patterns. Attendant representations of space involve the countryside as a diverse home accessible to all and not embossed with private property and policed by 'Keep Out' notices. Consequently, the representational spaces should involve a celebration of the local and the individually meaningful. Here, there is thus considerable potential to engage with the critical discourse on the rural idyll, as exemplified by Philo (1992) and Cloke (1993), which makes explicit the selectivity of that idyll for the general population, in respect of its gender, class, ethnic and other specificities and expectations.

The Scope for Spatial Transgression

Radical new structured coherences for the countryside are likely to have an extremely hard time establishing themselves **anywhere**, let alone nationally. This is because they represent an attempt to appropriate space 'creatively' rather than the efforts to 'divert' or appropriate space 'passively' which the other two alternatives represent (Lefebvre, 1991, pp. 167-168). The former involves the emergence of 'produced differences' within space, a creative process which:

> "presupposes the shattering of a system; it is born of an explosion; it emerges from the chasm opened up when a closed universe ruptures" (Lefebvre, 1991, p. 372).

In contrast, 'mere' re-appropriation concerns itself with 'induced' differences, which "remain ... within a set or system generated according to a particular law" (Lefebvre, 1991, p. 372). When this 'set or system' is seen as referring to capitalism, produced differences represent threats to the spatiality of capitalism itself.

Produced differences and efforts to build a 'counter-space' (Lefebvre, 1991, p. 349) tend to be treated harshly by the dominant system when they represent a significant threat to that system. Such a level of threat can be expected at this vulnerable time for rural structured coherences. The way in which such differences are suppressed are evocatively depicted in Lefebvre's work, possibly as a result of his own involvement with the ill-fated May 1968 'events' in France[5]. This suppression may involve violence:

"Abstract space ... reduces differences to induced differences: that is, to differences internally acceptable to a set of 'systems' which are planned as such, prefabricated as such - and which as such are completely redundant. To this reductive end no means is spared - not corruption, not terrorism, not constraint, not violence" (Lefebvre, 1991, p. 396).

For 'The Land is Ours' and similar campaigns, it may not yet have come to physical violence - although attention can be drawn to measures targeted at 'alternative' groups in the Criminal Justice and Public Order Act 1994 (Liberty, 1993; Sibley, 1994; Halfacree, 1996) - but more subtle efforts to bring these groups into line are apparent. In particular, there is the call by the government for such groups to abide by and work within the planning system as it stands. To do so would bring them back into the system, especially given that the planning system is a key guardian of the conventional urban-rural distinction.

Conclusion

In summary, new sets of structured coherences are forming in the British countryside in the wake of the demise of post-1945 agricultural productivism. Although simplified here in terms of super-productivism and the rural idyll, the detailed range of these coherences is likely to be substantial. Nevertheless, they are likely to conform to the commodified, privatised, segmented and class-ridden character that capitalist abstract space takes when materialised in place. The new rural structured coherences are thus unlikely to include radical spatialities, such as those suggested by 'The Land is Ours', without considerable physical, ideological and political struggle, since these spatialities are as much about abstract space as place:

"Place ... has the resources and capacity to transform space, but it cannot do so from the vantage point of place alone: political practices must ... be organised around place **in form** yet extend **in substance** to embrace space" (Merrifield, 1993, p. 527).

In this respect, the final words in this chapter will be left to Lefebvre (1991, p. 59):

"'Change life!' 'Change society!' These precepts mean nothing without the production of an appropriate space."

Notes

1. A version of this chapter was also presented at the 9th World Congress of Rural Sociology, University of Bucharest, Romania, 22-26 July 1996. Love and thanks to Irina.
2. 'The Land is Ours' have a detailed web site that explains their ideas and actions further: www.envirolink.org/orgs/tlio.
3. Whilst the idea of produced space is now reasonably familiar in the English-speaking world, thanks to the work of David Harvey and, most notably, Neil Smith's *Uneven Development* (1984), Lefebvre's original ideas are concentrated upon here.
4. Structured coherence represents an 'intermediate' concept within regulation theory (Jessop, 1990) in its emphasis on locally specific forms of regulation. This is in contrast to more overarching and possibly over-general concepts such as 'post-Fordism' (Cloke and Goodwin, 1992).
5. Consequently, and with relevance to the material discussed in this paper, Lefebvre pays some attention to the failure of various 'communal' experiments (for example, Lefebvre, 1991, pp. 53, 167-168, 379-381) to produce their own space and thus to be able to survive 'outside' the dominant system. A further and most apposite example of such failure would, of course, be that of the original Diggers (Hill, 1972).

Chapter 7

'Country Living': Rural Non-farm Population Growth in the Coastal Plain Region of North Carolina

Johnathan Bascom and Richard Gordon

Introduction

Prior to 1970, rural counties in the United States lost people to metropolitan ones. The process of rural out-migration, which began in the agricultural heartland of the Great Plains and Corn Belt, eventually spread to the Southern Coastal Plains and the Mississippi Delta (Johnson, 1985). During the early 1970s, however, this long-standing flow pattern reversed; migration to rural areas exceeded that of metropolitan ones (Tucker, 1976).[1] Nearly 80% of all non-metropolitan counties in the US gained population between 1970 and 1977, compared with only 47% during the previous decade (Johnson, 1985, p. 101).

Evidence was quick to show that this demographic reversal was not simply the product of urban spill-over. Phillips and Brunn (1978) and Garkovich (1989a) found net in-migration to all types of rural counties, ranging from the most remote rural ones to those neighbouring Standard Metropolitan Statistical Areas. And significantly, rapid growth occurred along highways and secondary roads outside incorporated cities and towns (Long, 1982, p. 1113). The re-population of rural areas continued through the 1980s and into the 1990s. The aim of this study is to examine factors that account for rural non-farm population growth among the largest agglomeration of rural counties East of the Appalachian Mountains – the eastern region of North Carolina.

Eastern Carolina

In 1990, North Carolina had a total population of 6.6 million people distributed among 100 counties. Over 2 million people live within the 41 counties that comprise the eastern portion of the state known as the Coastal Plain region. The remaining 4.5 million reside in the western regions of the Mountains and

Piedmont (Fig. 7.1). It was not until the 1970 census that any county in the Coastal Plain region had an urban area whose population was large enough to qualify for metropolitan status.[2] In 1990, only three counties had 'metropolitan' status leaving the remaining 38 counties classified as 'rural'.

The coastal plain of eastern North Carolina is a flat to gently undulating surface that lent itself to the formation of a central place system. Historically, towns and cities functioned as service centres for a predominantly agriculturally based economy.[3] Here, where agriculture and dispersed rural settlement has persisted for more than three centuries, a classic urban hierarchy exists, dominated by small towns and hamlets (Glade and Stillwell, 1986).[4] Only two limited access interstate highways (I-40 and I-95) run through the region. Most of the region still reflects a rural character and composition.

During the last five decades, the Coastal Plain region has experienced an overwhelming positive change in both the absolute number and proportion of rural non-farm residents.[5] From 1950 to 1990, only two counties suffered a net loss of rural non-farm residents; formerly rural areas adjacent to the expanding cities of Wilmington and Fayetteville were annexed, thereby decreasing the absolute number of non-farm residents in New Hanover and Cumberland counties. During the same 40 year span, only three metropolitan counties and one rural county experienced a negative change in the proportion of the rural non-farm residents (see Table 7.1).[6] The remaining 37 counties enjoyed a positive percentage change in the proportion of rural non-farm residents.

As seen already, counties are generally categorised as either metropolitan or rural. Therefore, to refine our analysis further, we subdivided the 38 rural counties into groupings based on the size of urban areas contained within them and their adjacency to metropolitan areas (Fig. 7.2). The four resultant categories are: metropolitan counties as containing a major city of 50,000 people or more (Metro); rural counties which are adjacent to metropolitan counties (RAM)[7]; rural counties with at least one major city of 25,000 to 50,000 inhabitants (RWC); and the remaining counties which, by definition, are not adjacent to any metropolitan counties nor contain any major cities and, thereby, are deemed most rural (Rural). These four groupings constitute the primary units of analysis for this study. Counties in the Rural category predominate in the northern portion of the Coastal Plain, counties in the RWC category lie in the central portion, and Metropolitan counties dominate the southern portion.

In the space of just two decades, the Metropolitan counties enjoyed an increase of 196,839 residents, virtually doubling their population by 1970. During the 1970s – a decade in which rural counties enjoyed their greatest growth – the pace of metropolitan growth slowed to 18% and remained virtually the same during the 1980s. Meanwhile, population growth in all three rural categories proved more variable. Between 1950 and 1990, the RWC counties had a total increase of 161,884 residents – growth of 50%. In keeping with the national pattern, counties in both the RAM and Rural categories suffered serious declines in the 1960s, then rebounded in the 1970s. By 1990, the RAM counties had a net increase over four decades of 133,577 – growth of 34%. The Rural group suffered from substantial out-migration during the 1950s and 1960s. After a decline of more than 6%, population growth revived. The total number of residents

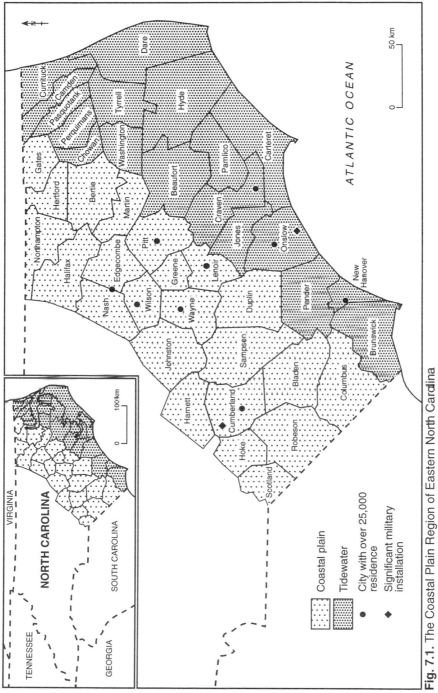

Fig. 7.1. The Coastal Plain Region of Eastern North Carolina

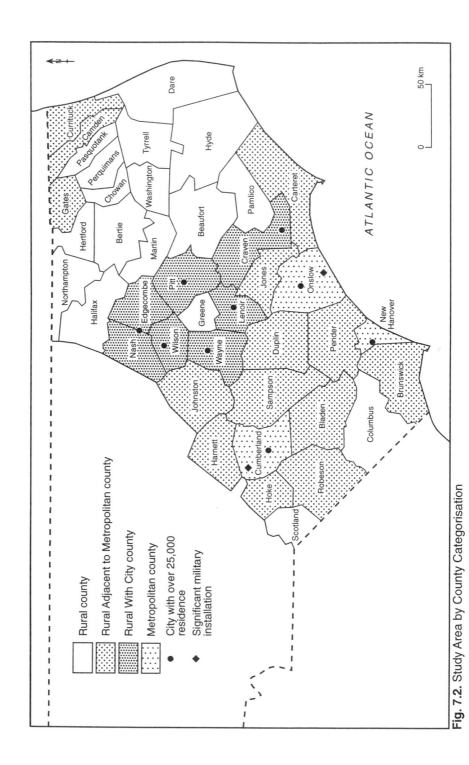

Fig. 7.2. Study Area by County Categorisation

Rural county

Rural Adjacent to Metropolitan county

Rural With City county

Metropolitan county

● City with over 25,000 residence

◆ Significant military installation

in the Rural group grew to 40,635 – a net gain of 16,594 – resulting in a modest gain of 5% over the four decades.

For the Rural county group, the proportion of rural non-farm inhabitants grew substantially in each of the last four decades (see Table 7.1). The percentage also increased in each decade for the RWC counties. Between 1950 and 1970, the proportion of rural non-farm residents in the RAM category increased dramatically from 40 to 71%. A slight decrease took place in the 1970s, but further growth followed thereafter. Since 1960, the percentage of non-farm residents fell in the Metropolitan county group due to the expansion of urban areas and subsequent reclassification. Metropolitan losses notwithstanding, the total percentage of rural non-farm residents in eastern North Carolina grew from 36% in 1950 to 55% in 1990, an increase in absolute numbers from less than 500,000 to more than 1.1 million.[8]

The Analysis

Reasons for non-metropolitan population growth are varied; explanations range from improvements in transportation infrastructure to industrial decentralisation to personal preferences for the quality of life afforded by rural areas. Next we examine several explanations in light of growth trends for the rural non-farm population in the Coastal Plain region.

Demise of the Farm Population

The Southeast is that region of the United States that has changed most dramatically during the later half of the 20th century. The shift from a largely agrarian society to a mixed economy generated substantial occupational changes among the region's rural populations. Figure 7.3 illustrates the rapidity of change within the agrarian economy of eastern North Carolina. By 1987, the agricultural sector employed less than 25,000 workers (Gordon, 1996, p. 61). Beginning in 1988, under a national amnesty programme, previously illegal migrant workers were allowed to register as citizens. This raised slightly the number of agricultural workers in the Coastal Plain region, but the percentage of rural farm residents continued to contract within all four county groupings. By 1990, rural farm residents consisted of less than 2.5% of the Coastal Plain population and the percentage continues to fall (Gordon, 1996, p. 77).

Agricultural mechanisation and the transition to corporate farming both reduced the number of rural residents who derived their income from agriculture. In North Carolina as a whole, the average farm size, measured in hectares per farm, has more than doubled since 1950 (Perrin and Sappie, 1990). Farm sizes are greatest in the eastern portion of the state. Other reasons for less rural farm residents include farmer retirement and out-migration, employment diversification, and occupational change. Currently 37% of 'farmers' in eastern North Carolina report their principal occupation as other than farming (Gordon, 1996, p. 77). The growth of industrial branch plants in rural areas outlying large urban centres, in small towns and in the countryside itself has allowed rural people to shift their occupation without changing their residential location.

Table 7.1. Rural non-farm residents in each county, 1950 to 1990 (percentages)

		1950	1960	1970	1980	1990
Rural	Beaufort	27.4	46.0	59.3	59.4	66.2
	Bertie	36.4	52.2	67.4	87.4	94.5
	Chowan	15.6	31.0	43.7	51.2	57.0
	Columbus	32.4	45.6	65.4	71.1	83.7
	Dare	98.6	97.7	70.1	51.7	81.2
	Duplin	34.3	50.4	59.8	66.0	76.6
	Gates	39.3	61.6	75.3	85.3	90.7
	Greene	17.4	38.1	59.0	78.0	92.9
	Halifax	30.1	36.8	47.2	59.4	59.5
	Hertford	35.8	38.0	50.9	63.3	66.5
	Hyde	58.4	69.9	82.2	83.9	92.3
	Martin	24.8	30.6	51.8	61.0	72.4
	Northampton	39.9	59.4	87.5	99.8	95.1
	Pamlico	71.3	83.5	76.8	87.6	98.4
	Pasquotank	34.3	37.1	41.1	43.0	52.1
	Perquimans	59.4	69.8	69.8	76.2	92.8
	Scotland	35.8	49.3	56.3	54.1	64.6
	Tyrell	56.2	68.1	74.7	89.0	94.1
	Washington	28.9	47.4	48.3	55.5	64.8
	Average	40.7	53.3	62.5	69.6	78.7
RWC	Craven	46.5	61.0	34.8	37.0	47.3
	Edgecombe	17.1	28.4	36.4	44.3	47.7
	Lenoir	20.0	33.1	43.7	45.6	48.1
	Nash	24.0	34.0	47.3	54.5	48.9
	Pitt	21.7	28.8	35.4	37.3	38.0
	Wayne	23.0	38.8	40.7	42.9	36.1
	Wilson	16.3	23.1	34.6	40.3	40.7
	Average	24.1	35.3	39.0	43.1	43.8
RAM	Bladen	40.6	58.1	76.4	77.9	82.1
	Brunswick	48.1	71.8	86.7	63.5	85.6
	Camden	57.0	73.1	91.9	87.2	96.8
	Carteret	52.2	67.6	67.7	54.5	75.6
	Currituck	53.7	81.2	63.6	58.2	97.3
	Harnett	30.2	46.1	58.6	57.1	73.5
	Hoke	37.3	45.7	60.2	61.2	82.6
	Johnston	28.7	38.7	53.4	56.9	71.4
	Jones	30.8	50.9	79.0	85.4	94.6
	Pender	41.8	68.7	80.4	76.0	97.2
	Robeson	35.2	37.2	52.3	53.5	75.5
	Sampson	23.8	37.4	54.4	63.9	76.4
	Average	40.0	56.4	71.2	66.3	84.1

Table 7.1. *contd*

		1950	1960	1970	1980	1990
Metro	Cumberland	44.3	46.0	21.3	12.7	12.9
	New Hanover	26.1	30.5	29.0	24.7	11.4
	Onslow	56.2	70.5	37.0	37.4	31.7
	Average	42.2	49.0	29.4	24.9	18.7

Notes: METRO - Metropolitan counties; RWC - Rural counties with city; RAM -
Rural counties adjacent to metropolitan areas (See text for definitions)
Source: North Carolina Census of Population, 1950, 1960, 1970, 1980, 1990

Non-metropolitan Industrial Growth

During the 1960s and early 1970s, rural areas – especially those of the South-East US – enjoyed a wave of industrialisation. Most of the growth occurred in non-metropolitan areas, reflecting a new location calculus that favoured rural locations (Long, 1982).[9] Improved transportation and dispersed markets had shifted the advantage towards peripheral areas in the US (Fisher and Mitchelson, 1981). Inexpensive land for factory sites, tax relief packages and other inducements offered by local government attracted industry further. As Lonsdale and Browning (1971, p. 260) observed, "The very character of Southern manufacturing – labour intensive, low profit margin, the achievement of labour economies often essential to maintain a competitive market position – tends to discourage many manufacturers from 'concentrating' in larger cities because of tighter labour markets and higher wages, despite some of the obvious advantages an urban location offers."

North Carolina typified the lack of a big-city orientation more than other southern states. Commenting on the industrial growth in the 1960s, Lonsdale and Browning (1971, p. 263) reported that 35% of all manufacturing in North Carolina occurred in places of less than 5,000 people compared with 25% for the South as a whole. The pool of low-wage workers made locating plants on the Coastal Plain region an attractive option. Lower taxes and large inexpensive tracts of land on which to locate were also incentives. But perhaps the key attraction for industry was this: North Carolina was the least unionised state in the nation.

The recession of the early 1970s hit manufacturing in the three Metropolitan counties of the Coastal Plain region particularly hard. Between 1970 and 1975, the percentage of the Metropolitan work force employed in manufacturing fell from 18 to 12%, and then continued falling. Meanwhile, industry with a rural orientation – the 'branch plant' phenomenon – continued growing among non-metropolitan counties of eastern North Carolina. In 1970, the percentage of the work force employed in manufacturing exceeded 20% for all three groups of rural counties and growth continued despite a pronounced recession. Prominent companies such as Dupont, Wyerhauser and Glaxo-Wellcome either established or expanded their operations during the 1970s. Although the percentage of the work force employed in manufacturing fell among all county groups during the 1980s, recent employment data show that the total number of industrial employees in the region continued to increase until the early 1990s (Bascom *et al.*, 1998).

83

Rural manufacturing continues to be an important source of off-farm incomes and rural non-farm population growth (RSSTF, 1993, p. 140). We should note that rural counties without major cities (Rural) have maintained a higher percentage of their work force in manufacturing throughout the last 25 years than have the other two rural categories – rural counties with small cities (RWC) or rural counties adjacent to major metropolitan areas (RAM) (Fig. 7.4). Although many of the declining labour force on farms shifted to occupations without changing their residential location during the 1970s, more recent journey-to-work data does not confirm the popular belief that commuting to industrial branch plants remains as important as it was during the 1970s.

Contraction of Commuting

The impact of nearly universal automobile ownership in permitting a longer journey-to-work for factory workers was not widely felt in the South until after World War II (Lonsdale and Browning, 1971, p. 262). During the decade of 1960 to 1970, the rural communities that grew fastest were, to a large extent, suburban residence centres (Davis and Clifford, 1973). This pattern still applies in eastern North Carolina. Among the four groupings, rural counties adjacent to metropolitan areas (RAM) had the highest percentage of their work force commuting over 30 minutes (29%) compared with 21% for the Rural group, 22% for the RWC group, and 15% for the Metropolitan group.

Distance to work is one indicator of the degree to which people choose to maintain their preference for homes in a rural environment rather than move closer to their place of employment. Table 7.2 combines journey-to-work data from the 1980 and 1990 North Carolina Census of the Population with that obtained from a large telephone survey conducted in 1994.[10] During the 1970s, commuting great distances from highway side homes may have allowed many East Carolinians to maintain their place of residence in rural areas (Hart and Chestang, 1978, p. 457). But what is most striking about the data presented in Table 7.2 is that commuting distances are markedly shorter in 1994 compared with those of 1990 and 1980. In all groupings, journeys of less than 10 minutes grew considerably. Apart from the lone exception of the RWC group, the percentage of longer journeys decreased. The shorter commuting distances of the rural population reflects two important new trends: work sites are shifting closer to rural residences; and the nature of the predominant form of employment for rural non-farm people is changing.

Strength of Service

The proportion of the total work force employed in the service sector has – in the case of all three rural groupings – increased substantially during the last two decades (Fig. 7.5). The highest percentage change took place in the Rural category – a 22% increase from 34% in 1970 to 56% in 1990. The RAM category had the next highest percentage change – an 18% increase to 71%. Rural counties with cities (RWC) experienced a 14% increase over the same twenty years; 45% of the total work force in 1990, up from 31% in 1970.

Table 7.2. Time of journey to work by county group in 1980, 1990, and 1994

	1980	1990	1994
RURAL			
Less than 10 minutes	26.1	47.1	25.9
10 to 19 minutes	22.2	50.3	27.5
30 minutes or more	36.2	42.8	21.0
RWC			
Less than 10 minutes	20.5	62.3	17.6
10 to 19 minutes	19.3	62.4	18.3
30 minutes or more	38.4	39.7	21.9
RAM			
Less than 10 minutes	19.6	45.2	35.6
10 to 19 minutes	15.6	47.5	36.9
30 minutes or more	31.6	39.8	28.8
METRO			
Less than 10 minutes	23.4	59.8	15.8
10 to 19 minutes	19.1	61.3	19.6
30 minutes or more	28.4	56.8	13.9

Source: North Carolina Census of Population, 1980, 1990; East Carolina
Survey Resources Laboratory Telephone Survey, 1994

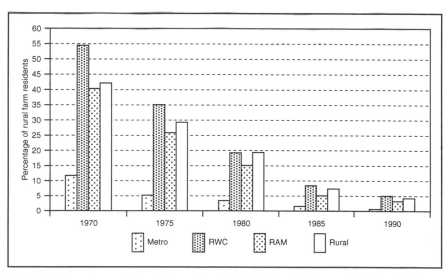

Fig. 7.3. Percentage of Rural Farm Residents by County Group, 1970 to 1990
(source: North Carolina Census of Population 1950, 1960, 1970, 1980, 1990)

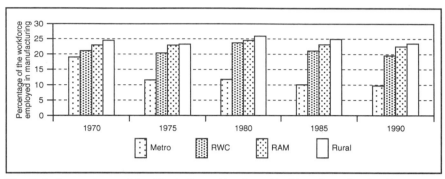

Fig. 7.4. Percentage of the Work Force Employed in Manufacturing by County Group, 1970 to 1990 (source: North Carolina Employment Security Commission)

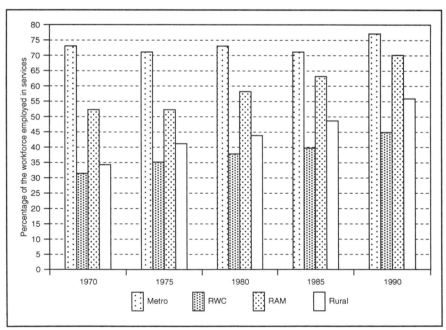

Fig. 7.5. Percentage of the Work Force Employed in Services by County Group, 1970 to 1990 (source: North Carolina Employment Security Commission)

The sources of service growth are many and varied. The region has two major state universities – East Carolina and UNC-Wilmington – as well as many smaller universities and community colleges. East Carolina's School of Medicine has fostered the growth of the medical and pharmaceutical industries in the area. For reasons of good politics, large tracts of land and an accommodating climate, eastern North Carolina has emerged as one of the nation's chief locales for military personnel (Glade and Stillwell, 1986, p. 128). Seven different military bases are located in the region and provide considerable employment for civilians.[11] The Coastal Plain region is also popular among retired military personnel who value moderate winter climate and ready access to base amenities like commissaries and lower cost medical care.

It is surprising to note that the Rural category outstrips that of the rural counties with cities (RWC) in terms of the percentage of service employment. Over half of that category's total work force now works in the service sector, which suggests that even the most peripheral counties have become service-based economies. In 1950, the largest percentage of the work force in the 17 Rural counties were employed in agriculture. By 1970, both manufacturing (24%) and service (34%) employed a greater proportion than agriculture (15%). The gap had widened further; by 1990 service (56%) compared with manufacturing (23%) and agriculture (5%).

The growth of a service-based, rural economy is strongest in the Tidewater sub-region of eastern North Carolina. Glade and Stillwell (1986, p. 108) reported "a population boom" in counties with "the Atlantic Ocean on their doorstep." Boosted by the rapid growth of coastal resort and retirement communities, the Tidewater area experienced the greatest population increase during the last decade (Skinner, 1992, p. 6). Between 1980 and 1990, the rural population of the Tidewater region grew by 18%. This compared with growth of 5% in the rest of the Coastal Plain region, 7% in the Piedmont, and 6% in the Mountain region. With the single exception of Hyde County, which does not have easily accessible coastal areas like its counterparts, all coastal counties enjoyed net in-migration.[12] The Atlantic coastline and its barrier islands are obvious attractions, but inland sounds, estuaries, and waterways offer another 7,360 km of coastal features that are increasingly sought out as prime recreational and retirement locales. Towns like Beaufort, Tarboro, New Bern and Washington have successfully attracted retirees and vacationers by rejuvenating their historical districts.

During the 1980s, nearly 40,000 people migrated into rural areas of the Tidewater sub-region (Skinner, 1992, p. 6).[13] Manufacturing jobs constitute less than 15% of total employment in the region. However, the infusion of large numbers of affluent and mobile retirees as well as out-of-state vacationers in search of recreation opportunities has strengthened and enhanced the demand for goods and services, generated new jobs and helped to curb the outflow of locals.

Conclusion

During the last four decades – 1950 to 1990 – the percentage of non-farm residents virtually doubled among the 38 rural counties within the eastern region of North Carolina. This study examined those factors that account for the dramatic increase of rural non-farm residents from 35% of the total population in 1950 to 70% in 1990.[14] Several 'lead' factors account for such growth. The first is the rapid decline of the rural farm and agriculture as sources of employment. In 1950, the rural farm population was substantial. The 41 counties in our study region were distributed in the following way: in one county more than 75% of its population were considered rural farm residents; in 15 counties the rural farm population ranged between 50 and 75%; in 19 counties it ranged between 25 and 50%; and, in only the six remaining counties did the rural farm population fall below 25%. But a precipitous decline occurred in the number and proportion of rural farm residents thereafter. By 1980, all 41 counties had less than 25% of their populations on rural farms. All this meant that many rural residents left for cities, thereby increasing the percentage of rural non-farm residents in the counties they left behind. Others continued to reside in rural areas, but secured non-agricultural sources of livelihood.

A dramatic increase in non-metropolitan manufacturing occurred during the 1960s and 1970s, especially in rural areas of southern states like North Carolina. Its effect was to shift agricultural workers into industrial employment with a concomitant proliferation of commuting and residential development in rural areas (Bunce, 1982, p. 171). A new pattern was established; given an opportunity to work in a nearby industrial plant, rural and small-town dwellers demonstrated a marked propensity to maintain their established place of residence and a willingness to commute great distances to work. Hence, industries began to draw their labour from remarkably broad geographical areas (Lonsdale, 1966, p. 115).

During the 1980s, two trends made substantial contributions to the growth of the rural non-farm population. First, rural residents extended the length of their journeys to work even further. Between 1980 and 1990, the percentage of the work force travelling for more 30 minutes grew in all four of our study categories. The second important trend was that of substantial in-migration to counties in the Tidewater portion of the study area. Strong growth in non-manufacturing employment occurred as a result of new residents, new development and increased tourism.

In the current decade, the importance of the service sector to rural areas has escalated. Commuting times have become markedly shorter, thereby marking the expansion of the service jobs in rural locations. The percentage of the total work force employed in service industries now exceeds 45% for RWC counties, 55% for Rural counties, and 70% for RAM counties. Thus, our study of eastern North Carolina suggests that future growth of the non-farm population will be more tightly tied than ever to the ongoing transformation of non-metropolitan areas into service-based economies.

Notes

1. During the 1970s, the rate of growth in urbanised areas was only marginally higher than rural places and that was due to rapid incorporation of rural areas into metropolitan ones.

2. The Federal Office of Management and Budget defines metropolitan counties as consisting of a central city with a population of 50,000 or more and any other contiguous counties that are metropolitan in character and are socially and economically integrated with the central city.

3. The region's leading agricultural commodities are tobacco, sweet potatoes, turkeys, hogs, broilers, cotton, peanuts, cucumbers, corn, soybeans and wheat.

4. Counties with medium-sized cities of 25,000 to 50,000 residents include Pitt (Greenville), Nash and Edgecombe (Rocky Mount), Wayne (Goldsboro), Wilson (Wilson), and Craven (New Bern and Havelock).

5. The last five censuses (1950 to 1990) defined rural non-farm residents as all persons living in areas not classified as urban that did not meet the following defined criteria for rural farm designation. For the 1950, 1960 and 1970 censuses, rural farm residents were identified as people living in urban areas as well as rural ones and, living on places of 4.1 ha or more from which sales of farm products amounted to $50 or more in the preceding calendar year, or, living on places of fewer than 4.1 ha from which sales of farm products amounted to $250 or more in the preceding calendar year. For the 1980 and 1990 censuses, rural farm residents were identified as those found only in rural areas, living on places of 0.4 ha or more from which at least $1,000 worth of agricultural products were sold during the previous year.

6. Dare County was the only non-metropolitan county which experienced a negative percentage change in the number of rural non-farm residents between 1950 and 1990. This coastal county had no areas designated as urban until 1990. Much of its rural non-farm population was reclassified as urban when the new urban designation was given.

7. Some of the metropolitan centres associated with RAM counties lie outside the eastern region (i.e. Raleigh and Durham as well as Myrtle Beach, South Carolina and Norfolk, Virginia).

8. In 1950, there were 493,849 rural non-farm residents among a total population of 1,372,157 residents in eastern North Carolina. In 1990, the absolute number had grown to 1,106,161 out of total population of 2,028,615.

9. Between 1967 and 1973 manufacturing employment in non-metropolitan areas rose by 11% while it declined by 3% in metropolitan areas (Fisher and Mitchelson, 1981, p. 304).

10. Each year the East Carolina Survey Resources Laboratory conducts a telephone survey of the 41 counties in our study area. A thousand residents were randomly selected for the 1994 survey. Each respondent was asked to identify the time required to make their daily journey to work based on the same time categories as were used in the Census of the Population.

11. The seven military bases are Ft. Bragg and Pope Air Force Base in Cumberland County, Cherry Point Air Station in Craven County, Camp Lejune and New River Air Station in Onslow County, Seymore Johnson Air Force Base in Wayne County, and Military Ocean Terminal in Brunswick County.

12. Coastal counties with the fastest growth rates are Brunswick (5% per year), Dare (5% per year), Curritcuk (4% per year), and Carteret (3% per year) (Glade and Stillwell, 1986, p. 108).

13. In comparison, the rest of the Coastal Plain suffered from out-migration

during the 1980s – a net migration decrease of 9,550 people in its rural areas and 12,372 from its urban centres.

14. These percentages represent the cumulative average derived by averaging the percentage for each of the 38 rural counties.

Chapter 8

Living in the Rural-Urban Fringe: Toward an Understanding of Life and Scale

Kenneth B. Beesley

Introduction

Living in a rural-urban fringe environment represents neither an urban nor a rural experience, yet it is both. The fringe, by its very character, allows residents to develop perceptions of life based on both the immediate residential environment and environments encountered during other life activities such as work or play. Despite this ambiguity (Bunce, 1981), the fringe is seen as more rural than urban by many of its inhabitants (Beesley, 1988a; Beesley and Macintosh, 1993). Part of this 'rurality' is associated with the fringe, even some of its more urbanised parts, being viewed as a relative oasis of tranquillity removed from the intense pressures of urban life. The rurality of the fringe is also linked to its recent past, frequently as an agricultural area, and the remnants of that past which persist in the landscapes of today, from barns, silos, pastures and cultivated fields to tractor dealers, seed companies and 4-H Club signs (Beesley *et al.*, 1996; Walker, 1987).

But all rural-urban fringe areas are not equal, despite the many similarities. Some fringes are proximate to metropolitan centres, allowing residents to gain advantages of metropolitan life such as a wide array of goods and services while enjoying the more 'rural' environs of the fringe, all at the cost of longer commuting distances in the greater metropolitan region. Other fringe zones surround small urban or non-metropolitan centres where the accessible goods, services and facilities are likely to be less diverse while the 'rural' is more readily available. Further, the fringe zone, metropolitan or non-metropolitan, can be characterised by variability in settlement form, from low density dispersed development through to a network of higher density compact nodal developments (Bryant *et al.*, 1982). Thus, rural-urban fringe areas demonstrably differ according to various objective criteria, but how do they compare when subjective dimensions of life and community are evaluated? Are perceptions of life and community, particularly satisfactions associated with these, different between metropolitan and non-metropolitan fringe environments? Alternatively, are the most important differences not between fringe areas, but between the fringe and the urban environment?

The aim of this paper is to examine results from a series of case studies which will help to address the questions posed above. These studies have formed a fundamental component of a research programme which has actively explored rural-urban fringe life and community satisfaction for over 15 years (Beesley, 1983a, 1983b, 1985, 1988a, 1991a, 1991b, 1993, 1994a; Beesley and Russwurm, 1989; Beesley and Bowles, 1991). Many documents based on findings from that research programme are available which develop individual studies or specifically focused research questions (Beesley, 1988b, 1994b, 1994c, 1995a, 1995b, 1995c, 1997; Beesley *et al.*, 1993; Beesley and Macintosh, 1994, 1995a, 1995b, 1993; Beesley *et al.*, 1995a, 1995b; Beesley and Walker, 1990a, 1990b; Coppack *et al*, 1990; Beesley and Macintosh, 1998), while this paper is designed to draw together several studies in a summary form, leading to a better understanding of fringe life and scale interrelationships.

Contextual Background

Case Studies from Southern Ontario

The southern Ontario case studies are outlined in Table 8.1, organised by their relative scale and locational attributes (Fig. 8.1 identifies key locations). The first division is into metropolitan and non-metropolitan studies, the former referring to work undertaken within Census metropolitan areas, as defined by Statistics Canada, i.e. areas where the total population in the urban core is greater than 100,000. The second division is whether the study is located in the urban part of the region or in the rural-urban fringe surrounding the urban area. The third division identified the geographical name of the local study area and when the case study took place.

Within metropolitan regions there are two studies in urban areas, in Metropolitan Toronto (1986-1987) and in the community of Bramalea (1990) a new town developed as part of the Municipality of Brampton North-West of Metropolitan Toronto but within the Toronto Census Metropolitan Area (CMA). Metropolitan fringe studies are from the Toronto region in two different time periods, 1984-1986 and 1992-1993, and from the fringe of the Kitchener-Waterloo CMA (1984). These metropolitan region studies represent 2,229 household interviews.

In non-metropolitan regions there is one urban study, in the City of Peterborough (1989). Three non-metropolitan rural-urban fringe studies are included, two from south-western Ontario around the cities of Guelph and Stratford (1984), and one from the Peterborough region (1990). Together these studies represent about 1,335 household interviews.

Fig. 8.1. Key Locations in Southern Ontario Case Studies

Table 8.1. Southern Ontario case studies

Urban Field Case Studies	Date	Sample Size
Metropolitan		
Urban		
Toronto	1986-87	284
Bramalea	1990	273
Rural-Urban Fringe		
Toronto		
1984-86	1986-87	609
1992-93	1990	688
Kitchener-Waterloo	1984	375
Non-metropolitan		
Urban		
Peterborough	1989	593
Rural-Urban Fringe		
Guelph	1984	281
Stratford	1984	229
Peterborough	1990	232
Total Sample Size		3564

Analytical Directions

Analytical directions taken are three-fold (see Table 8.2). Three blocks of variables are used: selected housing, community and socio-economic variables; life satisfaction measures; and community satisfaction measures. For each block of variables descriptive statistical results, either frequency distributions across categories or mean scores, are discussed. Satisfaction with one's life as a whole and satisfaction with one's community as a place to live are assessed using analysis of variance where the independent variables are length of residence, community attachment and age group. Finally, multiple regression models are used to identify important predictors of: satisfaction with life as a whole, where 11 domain satisfaction measures serve as the independent variables; and satisfaction with community as a place to live, where 18 community level concerns serve as the independent variables. In all cases the statistical findings are summarised for each of the case studies. More details on the statistical components of the research are available elsewhere (Beesley, 1995c).

Sample Characteristics: Housing, Community and Socio-economics

For each of the case studies background data on the characteristics of the sample are discussed briefly by considering the 15 variables identified in Table 8.2. Five variables deal with the respondents' housing: length of residence, tenure, period of construction, residence size (number of rooms) and residence condition. A second set of five variables focuses on community and include: prior community type, perceived quality of life in the current community compared with the prior community of residence, childhood community type, perceived quality of life in the current community compared with the community of childhood residence, and level of community attachment. The third set of variables offers insight into the socio-economic character of the samples and includes: age group, marital status, number of children, education completed and employment status.

Housing

Length of residence is an important characteristic in that it serves as a surrogate measure for familiarity with the area of residence, the assumption being that as length of residence increases so does knowledge of the local area and life in that area, at least from an experiential perspective. Among the urban case studies newcomers (resident for less than five years) account for 34-43% of respondents, while longer term residents (15 years or more) represent 26-32%. In fringe areas the non-metropolitan Guelph and Stratford regions are notable for proportionately fewer newcomers, at 22% and 17% respectively, with the Stratford fringe also characterised by a high proportion of longer term residents (42%). In the metropolitan Toronto fringe the separate samples from the 1980s and 1990s reflect the maturation of the fringe zone with an increase in the proportion of longer term residents from 19 to 28%.

Table 8.2. Analytical directions

Descriptive	Variables		Methods
Housing, Community and Socio-economics	Housing	Length of residence (years)	Frequencies
		Tenure	Frequencies
		Period of construction	
		Residence size (no. of rooms)	
		Residence condition	
	Community	Prior community type	
		Quality of life in present oommunity compared with prior community	
		Childhood community type	
		Quality of life in present community compared with childhood community	
		Community attachment	
	Socio-economics	Age group	
		Marital status	
		Number of children	
		Education completed	
		Employment status	
Life satisfaction	12 - Life as a whole and 11 domains		Means
Community satisfaction	19 - Community as a place to live and 18 community dimensions		Means
Selected variations			
Life satisfaction	Life as a whole by length of residence, community attachment and age		Analysis of variance
Community satisfaction	Community as a place to live by length of residence, community attachment and age		Analysis of variance
Satisfaction models	Domain to whole life model		Multiple regression
Community satisfaction	Community satisfaction model		Multiple regression

Revealed tenure patterns are consistent with expected patterns in urban and fringe areas in Canada. That is, home ownership is high, but higher in fringe zones. The 'new town' of Bramalea is identified as intermediate between the urban housing environments of Toronto and Peterborough and the rural-urban fringe locations. Residence age patterns also reflect differing settlement structures and development patterns. In mature urban environments 49-68% of homes date before 1961, and in the new town of Bramalea this was only 1%. In rural-urban fringe areas 20-55% of homes were constructed before 1961, with fewer in the Toronto fringe and more elsewhere. Newer homes, built since 1980, are more evident in the Peterborough urban study and in the 1992-1993 Toronto fringe case, both marking a relatively recent period of rapid development. Residence size is also an important measure, serving in part as a surrogate for wealth. In these southern Ontario case studies the Toronto metropolitan fringe stands out as a zone where there are relatively fewer modest sized (six rooms or less) homes (29-32%) and more larger (eight rooms or more) homes (49-52%).

Associated with residence age and size is the subjective evaluation of residence condition. In all cases most respondents describe their dwellings as in good or excellent condition, but this evaluation is markedly high in the metropolitan Toronto fringe at 82-85% of respondents. The non-metropolitan fringe of Peterborough, however, is identified by the lowest positive response (61%) linking it with residential environments that are slightly older and less affluent.

Community

Residential history, the sense of one's quality of life in the residential environment compared with other experiences, and feelings of community attachment all contribute to developing a better understanding of the differences between urban and fringe life, and between metropolitan and non-metropolitan life. The case studies reinforce the idea that not all fringe residents are urban escapees, rather many have come to the fringe from rural backgrounds (Beesley and Walker, 1990a). Indeed, in the fringe case studies 42-80% of the respondents grew up in a 'rural' environment (farm, non-farm, village or town), and 27-69% had lived in a rural community prior to their current residence. At the same time, 44-60% of urban respondents grew up in a city environment and 51-70% had lived in a city prior to their present residence area.

In general, most respondents consider their quality of life in their current community to be better or much better than in their childhood community, especially in the metropolitan Toronto fringe (62-66%). The new town of Bramalea stands out as the place where relatively many (16%) suggested that their quality of life compared with what was experienced in their childhood community was worse or much worse. In comparison to their prior community most respondents felt their quality of life was better or much better in their current community, again this is markedly so in the metropolitan Toronto fringe (67-69%). These findings support quality of life considerations as a genuine factor in residential mobility decision making (Jobes *et al.*, 1992).

Feelings of community attachment across most of the case studies are quite powerful, with over half of the respondents articulating a strong or very strong sense of community attachment. Only in Bramalea is this measure low, at 37%, with a higher response (13%) to the weak and very weak categories. The sense of community attachment is expected to play an influential role in discerning variations in satisfaction patterns, working as a link to both place and life.

Socio-economics

Age distributions help to describe the maturity of an area as well as its apparent attractiveness to differential age groups (Beesley, 1981; Bowles and Beesley, 1991; Dahms and Hallman, 1991). In most of the case studies the sample is characterised by respondents in the early to middle family ages of 25-44 years (40-49%), with the metropolitan Toronto fringe studies standing out with over 60% in this category. There are proportionately fewer older (55 years and over) respondents in Bramalea and the earlier Toronto fringe studies, while the non-metropolitan fringes of Guelph, Stratford and Peterborough all show more older respondents at about 33%.

The urban areas of Toronto and Peterborough are marked by lower proportions of married respondents (66%) and higher proportions of single (never married) respondents. In all other cases the married group is dominant, at 76-88%, with less than 16% single. This pattern is reflected in the number of children, with higher values in the 'none' category in the urban areas of Peterborough (26%) and Toronto (30%). However, all studies identify the ongoing importance of families with most respondents (54-68%) indicating two to four children in their family.

Educational attainment is another measure which serves in part as a surrogate of economic well-being, insofar as more education represents a better chance to achieve financial and material well-being (Lundy and Warme, 1986). The case studies are divided into three main educational groups. First, rural-urban fringe areas in south-western Ontario show higher rates of respondents with only an elementary school education (16-25%) and low proportions with a post-secondary education (24-34%). Secondly, there are areas characterised by higher levels of post-secondary education (57-61%), which include the City of Peterborough, Metropolitan Toronto, and the more recent (1992-1993) Toronto fringe study. In between are places with strong levels of post-secondary achievement (29-49%) and low levels of elementary school only attainment (4-11%).

Employment patterns reflect, to some extent, both age and education. In the south-western Ontario fringe studies the proportion of homemakers is higher at 26-35%. In all other areas the proportion of respondents who are employed or self-employed ranges from 57-76%, and in all cases except Bramalea, the retired represent a substantial group at 10-21%.

Life Satisfaction

Life satisfaction is assessed by considering: the mean satisfaction score for life as a whole and 11 major dimensions of everyday life, or life domains; variations in satisfaction with life as a whole by selected variables; and a satisfaction model. Satisfaction is measured, in all cases, using a scale where 1 equals 'completely dissatisfied', 4 equals 'neutral' or equally satisfied and dissatisfied, and 7 equals 'completely satisfied' (Beesley, 1985).

Patterns of life satisfaction are remarkably similar across all case studies (see Table 8.3). In every instance the mean scores for satisfaction with family life and friends are the highest, ranging from 5.6 to 6.1, while the mean scores for satisfaction with personal income and financial security are lowest, ranging from 4.7 to 5.2. Satisfaction with life as a whole is strong, with mean scores going from a low of 5.4 in Bramalea to a high of 5.9 in the metropolitan Toronto fringe. These patterns are consistent with the literature which suggests that satisfaction scores will be lower for those domains, such as income and finances, which relate to concerns that can be objectively measured, and higher for domains such as family life and friends where objective measures are unavailable (Andrews and Withey, 1976; Campbell, 1981; Pacione, 1982).

Table 8.3. Life satisfaction

Part 1	A	B	C	D	E
Satisfaction with:	Mean	Mean	Mean	Mean	Mean
Life as a whole	5.5	5.4	5.8	5.9	5.6
Standard of living	5.4	5.2	5.7	5.7	5.5
Financial security	5.1	4.8	5.4	5.2	5.1
Job	5.2	5.1	5.3	5.3	5.4
Personal income	5.0	4.7	5.1	5.0	5.0
Housing	5.5	5.0	5.7	5.7	5.7
Own health	5.6	5.4	5.9	5.9	5.7
Family life	5.8	5.6	6.1	6.1	6.0
Spare time activities	5.3	5.1	5.4	5.4	5.6
Friends	5.7	5.6	5.9	5.9	5.9
Independence	5.7	5.4	5.9	5.8	5.9
Part 2	F	G	H	I	
Satisfaction with:	Mean	Mean	Mean	Mean	
Life as a whole	5.6	5.5	5.7	5.8	
Standard of living	5,5	5.4	5.4	5.6	
Financial security	5.2	5.0	5.0	5.2	
Job	5.4	5.2	5.1	5.2	
Personal income	5.2	4.9	4.8	5.0	
Housing	5.8	5.7	5.6	5.8	
Local community	5.7	5.4	5.4	5.6	
Own health	5.8	5.8	5.7	5.9	
Family life	6.0	6.0	5.9	6.1	
Spare time activities	5.6	5.4	5.4	5.6	
Friends	5.9	5.9	5.9	6.0	
Independence	5.8	5.9	5.9	6.0	

Notes: A Metropolitan Toronto Total 1986-87; B Bramalea 1991; C Metropolitan Toronto Rural-Urban Fringe 1984-86; D Metropolitan Toronto Rural-Urban Fringe 1992-93; E Metropolitan Kitchener-Waterloo Rural-Urban Fringe 1984; F Non-metropolitan Guelph Rural-Urban Fringe 1984; G Non-metropolitan Stratford Rural-Urban Fringe 1984; H Non-metropolitan City of Peterborough 1989; I Non-metropolitan Peterborough Rural-Urban Fringe 1990

Analyses of variance test hypotheses that satisfaction with life as a whole will vary by selected housing, community and socio-economic criteria, and the results from the analyses are mixed (see Table 8.4). Length of residence emerges as a significant independent variable in only one case, the 1992-1993 Toronto fringe study, where satisfaction increases with length of residence. A stronger link between life satisfaction and community is found when community attachment is used as the independent variable. In virtually all cases, except the Kitchener-Waterloo metropolitan fringe, life satisfaction increases with the sense of community attachment. Age is revealed as a significant socio-economic measure in four cases: both Toronto fringe studies, the Guelph fringe and the City of Peterborough study. These findings are consistent with the literature in that life satisfaction tends to increase with age, a function of satisfaction of achievement and satisfaction of resignation with one's lot in life (Campbell, 1981).

Table 8.4. Selected life satisfaction variations (F statistics)

Part 1	A	B	C	D	E
Length of residence (years)	1.262	1.759	1.420	*3.788*	1.767
Community attachment	**6.057**	**5.783**	**3.496**	8.042	1.486
Age group (years)	1.539	1.386	2.928	5.670	1.161
Part 2	F	G	H	I	
Length of residence (years)	1.259	0.551	1.904	0.280	
Community attachment	*3.625*	**5.732**	**11.535**	9.101	
Age group (years)	2.438	0.602	*4.133*	1.881	

Notes: Probability: <u>0.05</u>; *0.01*; **0.001**
For explanation of A, B, C, ... I see Table 8.3

The domain to whole life model posits satisfaction with life as a whole as a function of the cumulative satisfactions associated with major life domains (Pacione, 1980, 1982). The results in these case studies are generally consistent with other research. The model serves to explain between 48% and 70% of the variability in satisfaction with life as a whole (see Table 8.5). In every case, satisfaction with one's standard of living emerges as a significant predictor, and in most cases (seven out of nine) satisfaction with one's family life is also important. An urban versus fringe variation is evident in that family life is the most important predictor in metropolitan Toronto and Bramalea urban studies, and in the City of Peterborough satisfaction with friends is ranked first. In all of the rural-urban fringe studies satisfaction with one's standard of living is the first ranked predictor. Also, in four of the fringe studies satisfaction with one's local community is a significant predictor, whereas this is the case in only one urban study, the City of Peterborough.

Community Satisfaction

Analyses of satisfaction with one's community as a place to live and various dimensions of that community reveal some interesting patterns. In urban studies in the Toronto CMA satisfaction with goods and services available is one of the higher mean scores (see Table 8.6). In all of the rural-urban fringe studies satisfaction with the community as a place to raise children is the highest mean score, with satisfaction with the people seen socially rated second. Relative

Table 8.5. Domain to whole life model (Beta statistics)

Part 1	A	B	C	D	E
	Mean	Mean	Mean	Mean	Mean
Standard of living	0.166	*0.213*	**0.345**	**0.329**	**0.270**
Financial security	0.096	*0.191*	0.039	-0.005	0.021
Job	**0.225**	*0.144*	0.086	0.070	0.106
Personal income	-0.016	-0.104	-0.082	0.080	0.009
Housing	0.077	0.050	0.006	*0.119*	0.110
Local community	0.068	0.004	*0.113*	0.023	0.121
Own health	-0.104	-0.067	0.003	0.074	*0.162*
Family life	**0.257**	**0.318**	**0.259**	**0.196**	0.003
Spare time activities	0.067	0.099	0.044	0.057	0.083
Friends	0.107	0.008	0.021	0.064	0.121
Independence	0.070	0.130	0.071	-0.008	0.036
Constant	0.949	0.956	0.592	0.577	0.279
R	0.731	0.756	0.693	0.695	0.780
R²	0.534	0.572	0.480	0.483	0.609
F	**25.854**	**28.141**	**42.643**	**45.636**	**40.270**

Part 2	F	G	H	I
	Mean	Mean	Mean	Mean
Standard of living	**0.428**	**0.370**	**0.194**	**0.277**
Financial security	-0.041	0.054	0.107	0.185
Job	0.037	0.065	0.073	0.028
Personal income	-0.001	0.007	-0.054	-0.029
Housing	-0.005	0.001	*0.144*	-0.050
Local community	0.036	*0.175*	0.081	**0.277**
Own health	0.066	0.001	**0.163**	0.024
Family life	**0.332**	0.041	*0.134*	**0.235**
Spare time activities	0.029	*0.192*	-0.043	-0.067
Friends	0.032	0.040	**0.207**	0.129
Independence	0.099	0.097	0.038	0.049
Constant	0.001	0.245	0.499	0.720
R	0.835	0.796	0.702	0.713
R²	0.696	0.633	0.492	0.509
F	**42.542**	**26.377**	**41.603**	**16.957**

Notes: Probability: 0.05; *0.01*; **0.001**
For explanation of A, B, C, ... I see Table 8.3

dissatisfaction with local taxes and local government activity is ubiquitous with mean scores ranging from 3.1 to 4.4. In four of the rural-urban fringe studies some dissatisfaction is also expressed for the services received for local taxes. Overall satisfaction with one's community as a place to live is strong but lower in urban areas, with mean scores from 5.1 to 5.6, and higher in fringe zones with means ranging from 5.7 to 6.0.

Variations in satisfaction with the community as a place to live are evident (see Table 8.7). Length of residence emerges as significant in three studies, both metropolitan Toronto fringe examples and the City of Peterborough, and community satisfaction tends to increase with length of residence. In all cases community attachment is a significant independent variable, as one would expect, with community satisfaction increasing with the sense of community attachment. Age serves as a significant criterion in four cases, both metropolitan Toronto fringe studies, the metropolitan Kitchener-Waterloo fringe and the City of Peterborough, and satisfaction again tends to increase with age.

Table 8.6. Community satisfaction

	A	B	C	D	E	F	G	H	I
Satisfaction with:									
Community as a place to live	5.6	5.1	5.8	6.0	6.0	5.8	5.9	5.5	5.7
Goods and services here	5.7	5.4	4.8	5.0	4.9	4.9	4.7	5.4	5.7
Ease of getting around	5.9	5.2	4.6	5.0	5.4	5.4	5.4	5.5	5.5
People in the community	5.4	4.8	5.6	5.8	5.8	5.7	5.7	5.3	4.9
Local government activity	4.4	4.0	4.2	3.6	4.3	4.4	4.2	4.1	3.5
Community as a place to raise children	5.4	5.1	6.0	6.0	6.1	6.1	6.0	5.6	5.9
People seen socially	5.6	5.4	5.7	5.8	5.9	5.8	5.9	5.8	5.7
Local taxes	4.1	3.6	3.9	3.1	3.6	4.1	4.5	4.1	3.4
Schools In the area	5.3	4.8	5.1	5.0	5.4	5.4	5.4	5.3	5.2
Police protection services	5.4	5.1	5.2	5.2	5.2	4.8	4.9	5.4	5.1
Fire protection services	5.6	5.4	5.6	5.5	5.8	5.4	5.5	5.6	5.2
Medical services and facilities	5.6	5.2	4.9	5.1	5.4	5.6	5.1	5.4	5.3
Welfare services	4.6	4.6	4.4	4.4	4.6	4.7	4.6	4.6	4.6
Waste disposal services	5.7	5.1	5.5	5.2	5.6	5.2	5.4	5.4	5.2
Road maintenance services	5.1	4.5	5.0	4.8	4.9	5.1	5.0	4.2	3.8
Recreational services and facilities	5.4	5.5	5.0	4.9	5.1	5.3	5.2	5.2	4.6
Cultural services and facilities	5.1	4.8	4.5	4.5	4.7	4.9	4.7	4.8	4.3
Overall quality of local services	5.5	5.1	5.0	5.0	5.2	5.2	5.2	5.3	4.9
Services for local taxes	4.8	4.3	4.3	3.7	4.1	4.4	4.5	4.5	3.7

Notes: For explanation of A, B, C, ... I see Table 8.3

Table 8.7. Selected community satisfaction variations (F statistics)

Part 1	A	B	C	D	E
Length of residence (years)	1.938	1.900	0.664	**4.784**	2.317
Community attachment	**22.703**	**33.860**	**44.674**	**31.891**	**14.478**
Age group (years)	1.539	1.386	2.928	**5.670**	1.161
Part 2	F	G	H	I	
Length of residence (years)	1.103	0.377	**6.258**	0.558	
Community attachment	**14.858**	**13.436**	**57.556**	**11.681**	
Age group (years)	1.740	1.874	**12.023**	0.919	

Notes: Probability: 0.05; *0.01*; **0.001**
For explanation of A, B, C, ... I see Table 8.3

The community satisfaction model suggests that satisfaction with community as a place to live is a function of the satisfactions with various community level concerns. In all case studies the model serves to explain between 57% and 68% of the variance in satisfaction with one's community as a place to live (see Table 8.8). The most frequent significant predictors among the nine case studies are satisfaction with: the community as a place to raise children (identified in 8 cases); people in the community (noted in 6 cases); and goods and services available in the community (also noted in 6 cases). In various rural-urban fringe case studies some predictors of community satisfaction are significant but negative contributors, notably satisfaction with fire protection services, medical services and facilities, recreation services and facilities, welfare services and services received for local taxes.

Discussion

It is clear that there are many satisfactions associated with life and community in urban and rural-urban fringe environments, in metropolitan and non-metropolitan regions. The case studies undertaken in various parts of southern Ontario over the 1984 to 1993 period reveal only minimal, and largely predictable, variations between urban and fringe environments in relation to the selected housing, community and socio-economic characteristics. Nevertheless, it is clear that the perceived relative quality of life in a community and the sense of attachment to it are important dimensions in residential choice and in community sustainability. There is also the postulate that satisfactions play an important role in decision making processes and in the long term well-being of a community.

Life satisfaction is associated strongly with both satisfaction with one's standard of living and with one's family life. It is in the rural-urban fringe studies, however, where satisfaction with the community as a place to raise children combines with a higher proportion of married respondents and fewer respondents with no children, reflecting a strong orientation to family life. In the rural-urban fringe there is the ability to trade-off the cost of commuting to an urban workplace, relatively fewer services and reduced access to goods and services,

102

with the benefits of more housing and property space, and a higher perceived quality of life. Furthermore, rural-urban fringe environments evidently receive residents from both urban and rural community backgrounds, offering the advantages of both while minimising the disadvantages.

Table 8.8. Community satisfaction model (Beta statistics)

	A	B	C	D	E	F	G	H	I
Goods & services	**0.30**	0.04	**0.17**	<u>0.13</u>	0.11	0.12	<u>0.17</u>	**0.52**	**0.65**
Ease of getting around	0.00	-0.01	-0.00	0.05	0.08	1.30	-0.08	0.06	-0.06
People in the community	**0.24**	**0.46**	**0.37**	0.05	**0.23**	**0.29**	**0.35**	0.01	0.08
Local government activity	0.00	-0.04	-0.01	0.03	0.05	0.03	0.07	0.01	-0.03
Community as a place to raise children	**0.33**	**0.30**	**0.27**	**0.42**	**0.26**	**0.27**	**0.27**	**0.16**	0.03
People seen socially	0.06	0.02	**0.17**	0.08	*0.17*	0.09	0.09	*0.10*	0.10
Local taxes	-0.05	-0.01	0.00	-0.00	0.03	0.06	0.08	<u>0.08</u>	0.09
Schools in the area	-0.02	<u>0.11</u>	0.03	0.09	0.10	-0.01	0.04	0.04	0.07
Police protection services	0.03	-0.12	0.03	0.08	0.00	-0.03	-0.07	-0.01	0.14
Fire protection services	0.02	-0.02	<u>-0.11</u>	0.06	0.02	0.03	-0.05	0.05	-0.13
Medical services & facilities	0.07	0.12	-0.06	**-0.19**	-0.05	0.04	0.14	-0.00	0.09
Welfare services	-0.05	0.09	-0.01	-0.04	0.01	-0.00	0.05	-0.02	*-0.19*
Waste disposal services	*-0.18*	0.02	0.03	0.01	0.05	-0.05	0.03	-0.00	0.09
Road maintenance services	-0.02	-0.03	-0.06	0.03	0.04	-0.01	-0.06	-0.01	-0.08
Recreational services & facilities	-0.09	-0.03	*0.15*	-0.06	<u>-0.13</u>	-0.06	0.03	0.01	-0.05
Cultural services & facilities	0.10	0.09	-0.08	-0.06	0.10	0.05	0.11	-0.01	-0.03
Overall quality of local services	0.15	0.07	<u>0.11</u>	0.07	0.10	0.11	-0.15	0.08	0.05
Services for local taxes	0.02	0.01	-0.09	0.02	<u>-0.16</u>	-0.11	0.04	-0.03	-0.07
Constant	0.42	0.01	0.71	2.22	0.66	0.53	0.12	0.08	1.51
R	0.79	0.82	0.75	0.60	0.78	0.76	0.78	0.79	0.78
R^2	0.63	0.68	0.57	0.36	0.60	0.58	0.61	0.63	0.60
F	**17.85**	**20.14**	**29.92**	**13.82**	**21.16**	**13.35**	**12.66**	**39.04**	**12.13**

Notes: Probability: <u>0.05</u>; *0.01*; **0.001**
For explanation of A, B, C, ... I see Table 8.3

This is not to say that the rural-urban fringe is an ideal environment for everyone. Indeed, problems are evident in relation to satisfaction with various local services. It is implicit in the research that the urban and fringe respondents are relatively 'successful' residents, i.e. they are dwelling in that area and have, in the main, moved to where they are from elsewhere. The research has not captured those who 'tried out' a fringe environment and left it for a more 'rural' or a more 'urban' setting

Conclusion

Dramatic metropolitan - non-metropolitan variations do not emerge, neither in the urban (Toronto or Bramalea compared with Peterborough), nor in the rural-urban fringe case studies. More evident are urban versus rural-urban fringe variations, but even these differences are at times small and subtle. These findings lead to conclusions, which are in part based on the evidence and in part on speculation.

1. Satisfaction with life is more a function of personal life circumstances than residential environment, whether urban or fringe, metropolitan or non-metropolitan. Though one's sense of community attachment is a strong factor, the nature of that influence does not vary between case studies.

2. Community satisfaction is strongly linked to the community as a social environment, a place to raise children and engage in social activities. While the range of goods and services available in urban areas may be greater, with a greater ease of accessibility to those goods and services, they are not an overwhelming force in community satisfaction.

3. The metropolitan fringe, particularly in the Toronto case, offers advantages which have helped to attract residents, such as more, better, or more readily accessible services, as well as the many opportunities associated with proximity to a major metropolitan centre. The non-metropolitan fringe offers other amenities, social and natural, which can be traded-off against travel to work in the local regional city.

4. The residential environments, in sum, exhibit a degree of synergy, where the whole is more than the sum of its parts. In the rural-urban fringe that 'whole' is strongly linked to one's standard of living and local community.

Acknowledgements

This work received financial support from the Social Sciences and Humanities Research Council of Canada, the Frost Centre for Canadian Studies and Development at Trent University and the Nova Scotia Agricultural College. I am grateful for that support. The work has also benefited from the contributions of Gerry Walker, York University, who collected the Metropolitan Toronto rural-urban fringe data, and Erica Donaldson, who collected the Bramalea data. I appreciate their help and permission to use their data. Finally, I acknowledge the contributions of the late Lorne Henry Russwurm, advisor and friend.

Chapter 9

Cluster Analysis of the Non-metropolitan Poor[1]

Bill Reimer

Introduction

Popular impressions of poverty in non-metropolitan Canada are equivocal. On the one hand it is invisible among the images of green fields, quaint villages, lakefront cottages, and country fairs offered by our media and memories. On the other, we hear of farm foreclosures, the disappearance of the fishing industry and trade wars in the forestry industry: all circumstances which contribute to poverty and deprivation. These two visions are not easy to reconcile, but we must try, since appropriate responses to poverty require an accurate view of the extent and nature of the problem.

The objective in this paper is to provide a description of the distribution of poverty in non-metropolitan Canada. Available data are used in an attempt to answer three questions:

- What are the characteristics of the non-metropolitan poor in Canada?
- Where are the various types of non-metropolitan poor located?
- What are the most promising directions for developing adequate explanations of non-metropolitan poverty in Canada?

Answers to these questions provide a structure for comparative analysis among existing studies of communities and regions in the country. A large part of the analysis is descriptive. As such, it means that the identification of the largest number of non-metropolitan poor, and the mapping of their distribution is paramount. This will form the basis for a subsequent more analytical examination of the non-metropolitan poor.

Theoretical Framework

Our analysis is exploratory, but it rests on theoretical perspectives which guide the selection of variables and techniques. Most of these perspectives are developed in the literature regarding poverty and inequality in general, although a few authors make explicit reference to differences between urban and non-metropolitan areas. Often this is discussed with respect to the impact which 'locality' has on inequality.

Theories of poverty and inequality can be organised in terms of three broad levels of analysis: individual or family centred, community or region centred, and institution or structure centred. In each case, the focus of analysis is on the major factors creating or maintaining economic inequality. The distinction between these three levels is not sharp, however, since many theorists integrate elements from each level and explanations at one level can often be used to complement explanations at another.

Individual or family-centred theories focus on the characteristics of individuals, their families, households, and sometimes associated culture to explain economic deprivation. In most cases, the emphasis is on the poor themselves, and the theoretical focus is on the way in which they fall into or fail to move out of poverty. Human capital theory is perhaps the most influential of these approaches (Rural Sociological Society Task Force, RSSTF, 1993). From this perspective, a person's level of income is assumed to be the result of the skills brought to the labour market. Those who do not have desired skills or are unwilling to market them are most likely to be among the poor.

In some cases, this argument is extended to include elements from cultural theories, which suggest that poverty is maintained through the internalisation of values which exacerbate the exit from poverty. In its most general form, the 'culture of poverty' perspective suggests that value and behavioural adaptations to a long history of poverty produce apathy, alienation, and community disorganisation which create greater barriers to overcoming poverty for the individuals involved (Lewis, 1986). A more specific form of this perspective attempts to explain a lack of entrepreneurial activity or a deficiency in the values and culture which encourage such activity (Gold, 1975). The tautological nature of these approaches has been noted in the critical literature (Ryan, 1976; Dolbearne and Dolbearne, 1976; Gotsch-Thomson, 1988).

Community or region-centred explanations for poverty focus on the characteristics of communities or regions as the sources for economic deprivation. As such they have direct relevance to the spatial characteristics of a rural focus. Some variations of the 'culture of poverty' theories as mentioned above imply a community-level value system which maintains its members in relative poverty. Community or regional poverty is thus considered to be the result of community values. A modern variation of this theme can be found in those theories which suggest that some communities are resistant to the type of changes which are necessary for adaptation to more general economic and social trends. The common element is that the community or region is treated as the central cause of economic deprivation, and thus is the focus of remedial action.

Institutional or structural explanations often begin with similar data to those above. The identification of causes is on general economic, political or social processes, however, not the local area response or lack of response to those processes. Rural poverty is seen to be the result of such things as capital accumulation (Gordon, 1972), labour market segmentation (Bluestone *et al.*, 1973; Edwards *et al.*, 1975; Kalleberg *et al.*, 1981), economic sector changes or 'restructuring' (Tomaskovic-Devey, 1988), economic cycles (Gans, 1972; Wachtel, 1971), political power (Baron and Bielby, 1980), market conditions (Rural Sociological Society Task Force, RSSTF, 1993), or some combination of these processes. In each case, non-metropolitan areas are disadvantaged by the trends in one or more ways.

It is not expected that the relative value of these various theories will be tested in a definitive manner. Rather, they will act as guides as when answering the three research questions. For example, the theories will provide direction when exploring the characteristics of the poor. Each theory implies that relatively high levels of poverty will be found among individuals, groups, or regions with particular characteristics. Those characteristics need to be included in the analysis. By using a wide variety of theoretical justifications for the variables, the chance that important variables are excluded is minimised.

Method

Cluster analysis is an appropriate technique to use for the first of the three research questions. It provides a way of classifying cases in terms of their similarity with respect to some set of characteristics. It also allows estimation of the relative number of cases which can be associated with each type, thus providing an indication of the relative vulnerability to poverty of various types of poor. When considering policy options, this latter information becomes important. The sensitivity of the technique is affected by two central aspects, however: the unit of analysis considered; and the variables selected.

As indicated above, there are three broad levels of analysis considered in the theoretical literature. It is extremely difficult to conduct empirical analysis which includes all levels at the same time since there are few data sets which contain them all. The usual approach is to focus on one empirical level and consider the ways in which it reflects theoretical processes at all levels. This is the way in which the present research proceeds: using the family or household as the basic unit of analysis.

The family or household has been selected for three main reasons. First, it represents a fundamental unit of economic interdependency in Canada. In most families there is a relatively high degree of sharing of resources, both economic and non-economic. As a result, the fortunes of one member are very likely to be visited on the others. Second, most of the theoretical processes relating to the emergence and maintenance of poverty have been discussed in terms of their implications for the family unit. Even theorists who emphasise structural or institutional processes are likely to develop the implications these processes have for poverty within families or households. Third, there are empirical data which are a close approximation to the type of economic interdependency at this level. A great deal of census and survey data are available which use the family, household, or economic family as the basic unit of representation.

The analytical results will also be affected by the variables considered when identifying the characteristics of poor families. The selection of variables might be limited in two major ways. The first is by the theoretical rationales considered. For this reason, a wide variety of theories regarding income in rural households are examined. The second way is through the data available for the analysis. This is a significant problem since the data available for national analysis is largely limited to census sources. These sources emphasise demographic and economic variables at the individual and household level, thereby creating a bias to these levels of analysis at the expense of local area or institutional information. There is little we can do to rectify this latter problem without extensive and expensive manipulation of both household and local area data. For this reason, we have chosen to proceed with more accessible data in spite of its limitations and integrate these limitations into our interpretation of the results.

The Census Public Use Microdata File on Households provides information on economic families[2] and unattached individuals for all of Canada. In addition, it includes variables which allow us to identify those families and individuals living in rural or small towns and those who fall below the low income cut-off as defined by Statistics Canada. Both of these characteristics are necessary in order for us to select the type of unit of interest. Families and individuals who live outside of Census Metropolitan and Census Agglomeration areas have been selected as the closest approximation to the non-metropolitan people which are the focus of interest. This includes those who live in smaller urban centres in Canada as well as those in rural areas.[3] The Statistics Canada low income cut-off has been used as the indicator for poverty in spite of the measurement problems which this entails. It is economic and somewhat absolute in its orientation, but it is by far the most convenient indicator of economic deprivation available in Canada.

The theories identified in the previous section have informed the choice of variables for the cluster analysis. The objective was to select variables representing each of the theoretical positions in order to reduce the bias to one or the other positions as much as possible. In many cases, this has meant rather indirect reflections of the theoretical positions because of the variables available on the microdata file.

Individual and family-centred theories suggest that those characteristics relating to the values and skills of individuals are likely to be related to poverty. The microdata file does not include any direct indicators of values, however, there are several characteristics related to the type of values identified in the culture of poverty literature. Family size, family structure and recent immigration status have been selected to reflect these characteristics. Human capital theories suggest that variables related to age, gender and education should be added to this list as indicators of the marketable skills which are available to the respondents.

Community or region-based theories are difficult to represent using family and individual level data. Community-based resistance to change and local employment opportunities are likely to be related to education and occupation which are also included in the analysis. The spatial manifestations of region-based processes will be examined by mapping the types of individuals and families once the cluster analysis is completed.

Institutional and structural explanations point to broad-based processes which disadvantage individuals and regions in various ways. Many of these explanations predict that these will be reflected in different levels of poverty in various economic sectors. The closest representation of this in the microdata file is the occupation variable. Class, labour-market and restructuring theories also imply that variation will occur with respect to the sources of income, the full or part-time organisation of work, and gender. Feminist theories reinforce the inclusion of gender as an important variable, both directly, and in conjunction with the extra burden which women bear for child care. A variable to indicate the presence of a single mother in the household has also been used.

In summary, the cluster analysis was designed to include the following variables:

* the number of children 0 to 17 years in the family
* the number of adults under 65 years in the family
* female single parent status
* an indication that both the household maintainer (HM) and spouse immigrated to Canada since 1977
* the average age of the HM and spouse
* the gender of the HM
* the average education of the HM and spouse
* the occupation of the HM (coded as dummy variables)[4]
* the major source of income (coded as dummy variables)[5]
* the average weeks of work for the HM and spouse

Results

Cluster analysis was conducted using the squared Euclidean distance measure of similarity. Since we were interested in being able to group clusters cumulatively, an hierarchical method was used, and since maximising similarity of cases within clusters was desirable, Ward's method for combining clusters was employed.[6]

Clustering methods leave a great deal of flexibility in the final identification of clusters. Ultimately, the decision must be based on criteria such as:

* how many clusters are theoretically appropriate for the phenomenon being investigated,
* how many clusters can be conceptualised and manipulated, and
* the points where there are major changes in the similarity or dissimilarity between clusters.

The 15 largest clusters were selected for more detailed analysis. This number provides sensitivity to regional variations while reflecting differences in the ways that poverty is established and maintained. In addition, the number of clusters is small enough to be conceptually manageable.

Once the 15 clusters are identified by the software it is still necessary to provide theoretical meaning to the groups chosen. This is achieved by examining the means of the standardised scores for each of the variables used. These values are provided in Table 9.1 for the six largest clusters. A convenient way to identify the cluster types is in terms of the variables with the largest means (positive or negative) for that cluster. For example, cluster 1 is characterised by large values on the standardised scores for the average number of weeks worked (-0.74), net government income (0.54) and average age (0.53). The negative value for the number of weeks worked indicates that families in this cluster are likely to include those who are employed part-time or are not in the labour force. Combined with the high values for government income and age, the families in this cluster are characterised as elderly, largely unemployed and receiving government payments, probably in the form of retirement payments. An examination of some of the other means for this cluster indicates that the household maintainer is likely to be male (-0.51), unlikely to have a primary occupation (-0.51), and the average family education is likely to be low (-0.46).

Table 9.1. Standardised means for first 6 clusters

Cluster:	1	2	3	4	5	6
No. children	-0.29	0.03	-0.12	0.31	-0.08	0.24
Single mother=1	-0.45	-0.47	**1.39**	-0.47	**0.55**	-0.28
Female HM=1	-0.51	**-0.57**	1.64	**-0.60**	0.76	**-0.35**
Age of HM and spouse	**0.53**	-0.07	0.14	-0.33	-0.34	**-0.43**
Family employment income	-0.22	0.44	-0.39	0.41	0.12	0.33
No. adults <65	0.28	0.23	-0.66	0.23	-0.28	0.02
Family net income	-0.08	-0.08	-0.11	-0.08	-0.05	-0.08
Family gov't income	**0.54**	-0.36	0.43	-0.10	-0.13	-0.23
Family other income	-0.16	-0.19	-0.15	-0.19	-0.07	-0.18
Avg. educ. of HM and spouse	-0.46	0.06	-0.35	0.03	0.18	0.32
Both HM and spouse immigrants=1	-0.10	-0.10	-0.10	-0.10	-0.10	-0.10
Avg. weeks worked by HM and spouse	**-0.74**	**0.67**	**-0.89**	0.08	**0.41**	**0.38**
Professional HM=1	-0.24	-0.24	-0.24	-0.24	-0.24	-0.24
Artistic HM=1	-0.08	-0.08	-0.08	-0.08	-0.08	-0.08
Clerical HM=1	-0.17	-0.17	-0.17	-0.17	-0.17	-0.17
Sales HM=1	-0.19	-0.19	-0.19	-0.19	-0.19	-0.19
Service HM=1	-0.29	-0.29	-0.29	-0.29	**3.48**	-0.29
Primary HM=1	-0.51	**1.93**	-0.43	-0.52	-0.52	-0.52
Processing HM=1	-0.20	-0.20	-0.20	-0.20	-0.20	-0.20
Manufacturing HM=1	-0.27	-0.27	-0.27	-0.27	-0.27	**3.69**
Construction HM=1	-0.31	-0.31	-0.30	**3.19**	-0.31	-0.31
Transportation HM=1	-0.19	-0.19	-0.19	-0.19	-0.19	-0.19

Using the same approach, the largest 15 clusters are characterised in the following way:

Cluster 1: non-employed, government income dependent, elderly
Cluster 2: full-time primary workers
Cluster 3: single mothers
Cluster 4: construction workers, predominantly male HMs
Cluster 5: service workers, predominantly female HMs
Cluster 6: manufacturing workers, predominantly young males
Cluster 7: professional, educated, full-time HMs
Cluster 8: young families with the HM in processing occupations
Cluster 9: families with little employment income but high net income, mainly primary workers
Cluster 10: HMs in transportation, with young children
Cluster 11: HMs in sales, full-time
Cluster 12: families receiving investment income, non-employed, elderly
Cluster 13: clerical, women HMs, single mothers, educated
Cluster 14: immigrant families with young children, predominantly sales HMs
Cluster 15: HMs in arts, educated

It should be apparent from Table 9.1 that although families have been classified into mutually exclusive clusters, there is considerable conceptual overlap in the variable loadings on each cluster. For example, clusters 1 and 2 both load in a high negative fashion with respect to the gender variable, indicating that families in these clusters are likely to have a male as the household maintainer. However, they differ with respect to several of the other variables that are highly-loaded such as age and the average number of weeks worked. The software used to undertake the clustering sorts out these similarities and differences in an hierarchical fashion. Thus clusters are cumulatively combined until all of the cases are included in one large cluster.

Figure 9.1 provides a representation of the way in which the last 15 clusters would be combined by this method. For example, of the 15 clusters, clusters 2 and 9 are identified as the most similar. They would be combined on the first step. The next most similar clusters are 1 and 12. Once these latter two are combined, the next step would combine them with cluster 3, and so on.

In general, the 15 clusters fall into two broad groups: those in which people are unemployed; and those where they are employed. Unemployed clusters are of three major types: the elderly, young, single mothers, and a few primary workers with non-employment sources of income. Of those employed, there are several different clusters representing a variety of occupations: primary, construction, service, manufacturing, professional and processing sectors being the largest. The occupations show considerable correlation with the gender of the household maintainer.

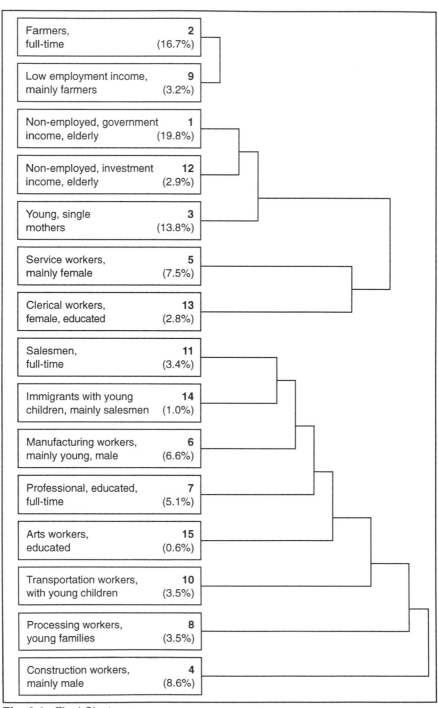

Fig. 9.1. Final Clusters

Mapping the Clusters

In order to answer the second research question it is necessary to match the information regarding the clusters to appropriate geographical units. The number of families and individuals in the Public Use Microdata File was too small to provide adequate representation for important regional differences. As an alternative, the primary characteristics of the clusters were matched with the more detailed information from the User Summary Tapes of the 1986 Census[7] to prepare a geographical profile. The distributions for the first four clusters were prepared using this data.

The Census Division (CD) is the most convenient unit for the geographical representation. Figure 9.2 illustrates that the CDs with the highest proportion of rural households below the Statistics Canada low income cut-off are located in Newfoundland, sections of Nova Scotia, New Brunswick, Québec, Central Manitoba, Saskatchewan, Northern Alberta, and in one CD in British Columbia.[8] Ontario, southern Alberta, British Columbia and the North have particularly few low income households as would be expected from the economic advantages of these regions. However, this general data masks some important variations in the type of households. These are illustrated by an examination of the distribution by cluster.

An estimate of the proportion of each cluster type was made for each CD and these estimates were plotted in a thematic style. It was not a straightforward procedure, however, since direct analysis would have been extremely costly. Instead, the estimates were constructed in an indirect fashion from enumeration area data.[9] An index was constructed for each enumeration area (EA) using the major characteristics of each cluster. For example, the index for cluster 1 was constructed by calculating the percentage of people over 65, the percentage of people who do not work full-time and the percentage of families in the bottom 20% of the income scale in each non-metropolitan EA. These values were then weighted by their mean score from the cluster analysis.[10] Finally, the cut-off score for the top quintile was calculated for all EAs in Canada and a score for each CD was calculated as the percentage of non-metropolitan EAs in the CD which was higher than this cut-off.

It was necessary to employ a rather indirect technique for clusters 1 and 3, but for clusters 2 and 4, a more direct measure was calculated. For these clusters, CD-level data regarding the occupation of the HM in non-metropolitan, low income households were available. Since the type of occupation played such a prominent part in the definition of these clusters they were used as indicators. The results of this analysis can be found in Figs 9.3 to 9.6.

The largest cluster of low income households is represented by an index for rural elderly who were unlikely to be employed full-time (see Fig. 9.3). The distribution of these households parallels closely the overall distribution of low income households in the Atlantic region, but it varies for other regions. They are over-represented among low income households in several regions of Ontario, Northern Manitoba and Saskatchewan, and several locations in British Columbia.

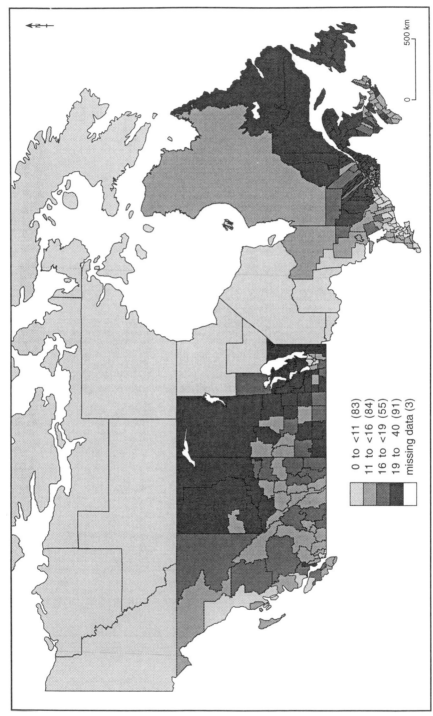

Fig. 9.2. Percentage of Rural Households below Statistics Canada Low Income Cut-off (Canadian Census Divisions)

Legend:
- 0 to <11 (83)
- 11 to <16 (84)
- 16 to <19 (55)
- 19 to 40 (91)
- missing data (3)

500 km

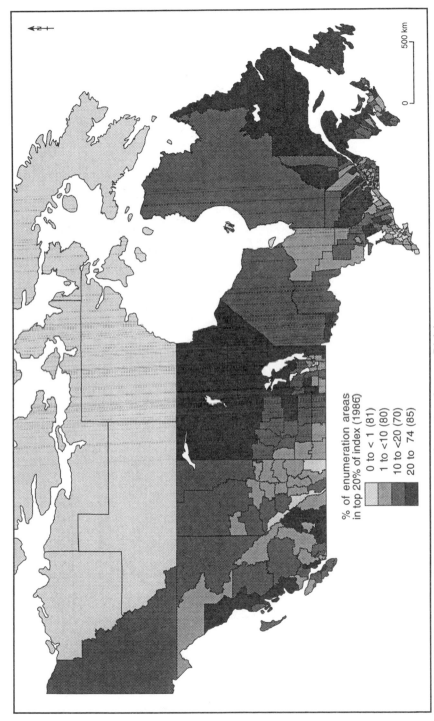

Fig. 9.3. Index for Cluster 1 Type Households in Canadian Census Divisions

% of enumeration areas
in top 20% of index (1986)

0 to < 1 (81)
1 to <10 (80)
10 to <20 (70)
20 to 74 (85)

500 km

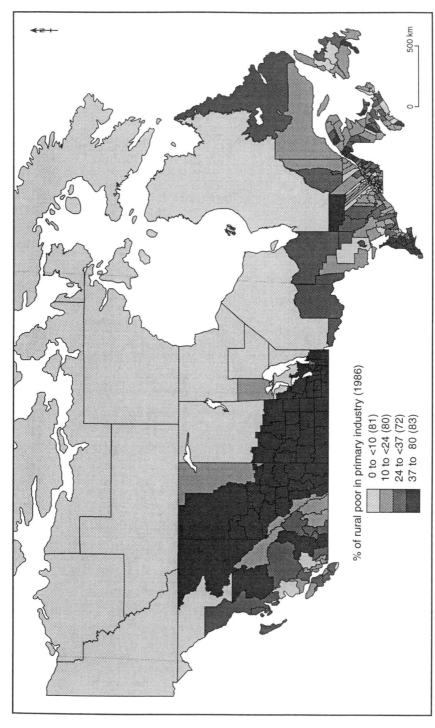

Fig. 9.4. Rural Low Income Households Working in Primary Industrial Sector

% of rural poor in primary industry (1986)

0 to <10 (81)
10 to <24 (80)
24 to <37 (72)
37 to 80 (83)

500 km

A common characteristic of these regions is their remoteness from major urban centres. This is most likely a reflection of the age-dependent patterns of out-migration identified by demographers (Lowry, 1966).

The second cluster has a very different distribution (Fig. 9.4). This represents the percentage of poor rural households in which the household maintainer is employed in a primary industry (agriculture, fishing, forestry or mining). The most striking feature of this map is the extent to which farming areas are over-represented. This includes the southern regions of the prairies, the Peace River district, and several more isolated regions in British Columbia, southern Ontario, Québec and the Atlantic provinces. However, an appropriate interpretation of these results requires a more detailed analysis, since the low income status is likely to be a combination of many factors: the relatively low economic viability of some farms, the low wage structure in farming and the use of farm status as a basis for income tax advantages.

Rural single mothers with low incomes is the third cluster identified by the analysis (Fig. 9.5). In this case, the highest proportions are found in the more remote regions of the country, with the exception of the North, Labrador, the Gaspé, mid-Québec, northern Ontario and the prairies, the South-East of British Columbia, the remote coastal communities and the Yukon all have high proportions of poor single mothers. The reasons for this are unclear. Further analysis will look at the patterns of mobility, marital status, household structure and the role of aboriginal communities.

The final cluster to be mapped is that of poor construction workers in rural regions (Fig. 9.6). This distribution shows the same tendency to be high in relatively remote regions as found with single mothers, but the specific regions are different. As expected, the levels are low in the North and near to most large urban centres, but the characteristics of those areas where the levels are high will require more analysis before any further patterns become apparent.

Conclusions

This analysis confirms what others have shown in various ways: the poor are a diverse group. This is the case, not only for those who live in metropolitan areas, but for small towns and rural regions as well. In addition, the relative frequency with which the various types of families occur has been illustrated. The largest group to be identified is the elderly, most often unemployed and dependent on government assistance. This assistance has not been enough to bring them over the low income cut-off. Examples of these families can be found in most parts of Canada. However, they seem to be most concentrated in the Atlantic region, Québec and more remote parts of most Western provinces. Further investigation should concentrate on their access to services in those areas, particularly since the elderly are often limited in their transportation options.

The second largest group contains those families in which the household maintainer is in a primary occupation. From the geographical distribution, most of these appear to be farm families, since they are concentrated in the major farming

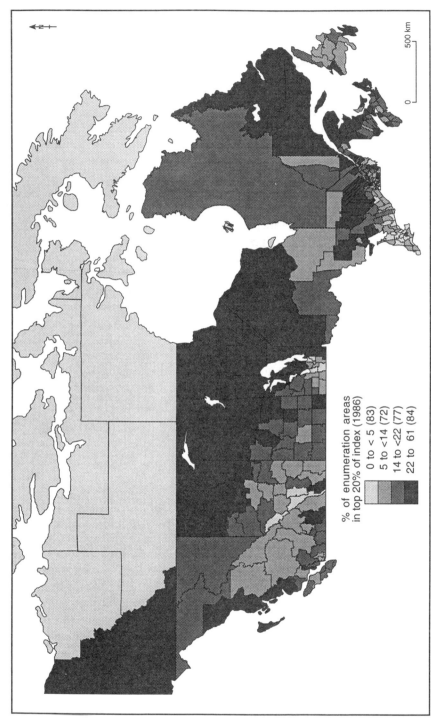

Fig. 9.5. Index for Cluster 2 Type Households in Canadian Census Divisions

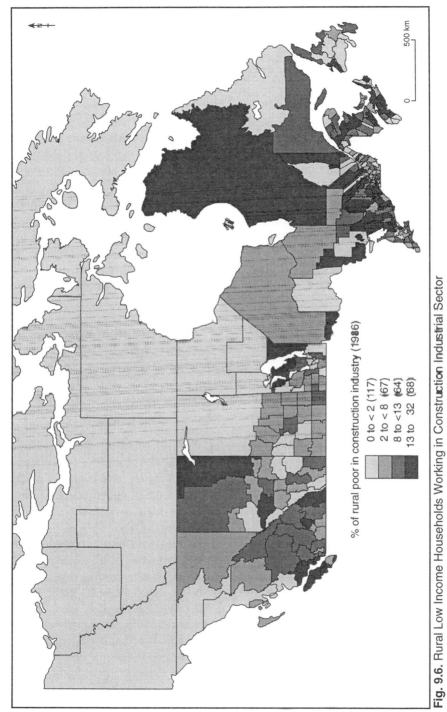

Fig. 9.6. Rural Low Income Households Working in Construction Industrial Sector

% of rural poor in construction industry (1986)

0 to < 2 (117)
2 to < 8 (67)
8 to <13 (64)
13 to 32 (68)

500 km

regions of the country. Use of income as an indicator of economic deprivation is likely to be particularly problematic with this group, since there is a much closer integration of commercial and domestic finances in most farming operations than will be found in families oriented to salary income. As a result, some farms may be used as a form of tax shield. The implications of this difference for deprivation issues requires additional research.

The third largest cluster is a special problem for the northern parts of most provinces. Single mothers are especially vulnerable to low income, which is as significant in non-metropolitan areas as metropolitan ones. In addition, the number in Québec is over three times that of Ontario, the next largest province. Further research is required to investigate the social situation of these women, especially with respect to the support networks which are available to them. Access to transportation and child care are likely to be crucial factors in their ability to survive and move out of poverty.

The fourth cluster represents one component of the rural working poor. Their overrepresentation is highest in the more remote regions of the country, but the reasons for this are unclear. Their relatively high level of dependence on employment income, high number of children and working-age status suggests they may be vulnerable to low-wage work and heavy family demands. Should this be the case, further research is crucial before implementing policy regarding work structures, child welfare or family structures.

An examination of those variables which play the largest part in differentiating clusters provides few hints regarding the relative strength of the theories upon which they are based. Employment and economic sector variables play the most prominent role along with age and gender. Most of the theoretical positions predict this, however, with the possible exception of the region-based theories. This latter perspective is supported by the spatial analysis. None of the theories can be eliminated on the basis of the cluster analysis.

Further research requires a more detailed study of each of the types identified by the cluster analysis. At the same time, we can move to more theoretically sensitive types of analysis in order to examine not only the characteristics of the groups, but the processes which are likely to produce and/or maintain their disadvantaged position. There is no reason to expect that they will be the same for each of the types identified.

There are a number of policy implications suggested by these results. The significant differences in the types of clusters identified and the relatively large numbers of households represented by each group suggest that programmes directed to non-metropolitan poverty must be multi-faceted. Broad-based programmes, which are not directed to specific groups, are likely to have limited success since the types of households among the non-metropolitan poor are highly disparate in both their social and spatial circumstances. The implementation of old age security has demonstrated that programmes directed to specific groups can be extremely effective. Such measures may equally apply to the other types of households which have been identified.

Notes

1. This research is supported by a grant from the Social Science and Humanities Research Council of Canada. I would like to thank Shawn Berry, Barry Ellison, Jasper Moiseiwitsch, Tom Saldanha, Elizabeth Szekely and Grace Young for their help in the formulation of ideas and analysis of data associated with the research.

2. "An *economic family* ... includes all persons related by blood, marriage or adoption living in the same dwelling." (Statistics Canada, 1988, p. 64).

3. Non-CMA/CA areas are defined by Statistics Canada as those outside census metropolitan areas or census agglomerations. They may be urban (an area having a population concentration of 1,000 or more and a population density of 400 or more per square kilometre), or rural (areas outside of urban areas) (Statistics Canada, 1988, p. 76).

4. Fifteen occupation groups were collapsed into eight: managerial, administrative and professional occupations; artistic, literary, recreational and related occupations; clerical, sales and service occupations; primary sector occupations; processing occupations; machine production occupations; construction; and transportation occupations. The ninth (other occupations) was excluded from the analysis to avoid collinearity problems.

5. Five sources of income were identified: no income; wages and salaries; self-employment; government transfers; and investment income. The sixth category (other income) was excluded to avoid problems of collinearity.

6. Ward's method produces results with the minimum variances within the clusters. Given the exploratory nature of the research, a contiguity constraint was not included in the cluster analysis (Abler *et al.*, 1971).

7. These tapes do not include information for some First Nations reserves. This must be kept in mind when interpreting the results from rural areas.

8. It is important to remember that the structure of Canadian CDs for 1986 means that some of the large areas represented are heavily influenced by a few large communities (usually in the southern part of the CD).

9. The enumeration area rather than the CD was chosen for this step of the process in order to minimise the ecological bias which might occur by using the larger unit of analysis.

10. For cluster 1, this index was calculated using the following formula:
((% not working full-time) x 0.74) + ((% over 65) x 0.53) + (% in bottom 20% of income).

Chapter 10

Deindustrialisation and Rural Economic Restructuring in Southern West Virginia

Tyrel G. Moore

Introduction

Resource-based economies underlie an important class of rural systems which occur at regional scales in the United States. As raw material suppliers, these rural regions and their economies are linked dynamically with the performance of urban industrial economies. Appalachia's regional economy was integrated into the national economy as spread effects emanated from 19th century industrial expansion in the North-East and Midwest. Growth in those areas created increased resource demands and eventually connected the more remote, but highly valuable, coal fields in eastern Kentucky and West Virginia to urban industrial areas (Moore, 1998).

Mining, railroads and manufacturing formed the pillars of this 19th century structural shift in Appalachia. In the long term, all of these proved to be highly sensitive to technological changes and volatile business cycles (Newman, 1972). These developments brought rural Appalachia into the national economic mainstream, but were shaped in a narrowly defined scope that built local economies lacking diversity and depending heavily on external economic and decision-making influences. By the 1960s, a long history of dependence on outside demands and a sporadically performing coal industry placed Appalachia at the centre of federal concern for economically depressed areas. It had become a region filled with small agricultural market towns with no markets, declining rail centres and declining mill and mining towns (Newman, 1972). Southern West Virginia was especially vulnerable because the development of its high-quality metallurgical coal had been influenced almost wholly by outside economic decisions.

An activist literature has argued that the economic and social inequity produced by this economic development system represents a form of colonialism which relegated the Central Appalachian coal mining areas of Kentucky, Virginia and West Virginia to a Third World socio-economic status in America (Lewis, 1991). Geographically more compelling interpretations have been offered which identify Appalachia as a periphery of urban-industrial core areas of the North-East and Midwest (Walls, 1978; Moore, 1994). The colonial development and core-periphery models theoretically agree in the sense that they illustrate rural Appalachia's dependence on and sensitivity to industrial linkages with heavy manufacturing areas to the North.

This paper focuses on economic change in southern West Virginia during the 1980s. In that decade, the area was profoundly affected by national economic restructuring and accompanying deindustrialisation in its traditional extra-regional markets. Much has been written on the impacts of deindustrialisation in urban industrial areas but less attention had been devoted to the effects on rural places. These have been alluded to rather broadly (Gaventa, 1987; Jakle and Wilson, 1992; Couto, 1994) and their specific connections with landscape and population dynamics in southern West Virginia's mining towns have been documented (Moore, 1998). This research extends those discussions to point out that while service sector employment absorbed industrial job losses and became a growth engine of the American economy, no parallel trend occurred in southern West Virginia's mining counties. Much of that failure to restructure can be attributed to development patterns that accompanied the area's singular reliance on the coal industry. The significance of that industry, and especially its dramatic collapse in the 1980s, must be placed in regional and national contexts. Backgrounds on the expansion and technological change of the region's coal industry are presented first, followed by discussions of deindustrialisation at national and regional scales. Finally, comments are offered on three post-industrial developments that symbolise the area's economic stagnation: the area became the target of attempts to use abandoned mining sites as landfills for out-of-state waste; externally initiated religiously-affiliated relief 'missions' have been organised toward improving the quality of life for the area's residents; and travel agencies far outside the region promoted the area as a destination for 'third world poverty' tours. Field interviews, regional media coverage and data from the West Virginia Department of Employment Security and the United States Bureau of the Census are used to develop the discussion.

Study Area

McDowell County, West Virginia, and adjacent Wyoming County, are the geographical focus of the study. These two counties were the centre of the area's mining activity and productivity from the extensive and thick Pocahontas coal seams. To the north, Raleigh County was underlain by valued multi-purpose coals of the Flat Top Mountain production field (Pocahontas Fuel Co., 1936). Beckley emerged as the headquarters and service centre for mining operations in that field. On the eastern edge of the coalfields, Bluefield and Princeton held railway repair shops, mining administrative offices and functioned as a shipment point to eastern seaboard ports. Welch evolved similar roles as the north-western gateway to industrial linkages in the Midwest (Moore, 1998). Outside these urban places, related mining activities tied these counties together as contiguous parts of southern West Virginia's past and contemporary economic conditions. Mining was not a sustained urban-forming process. In 1990, Wyoming County was 100% rural, McDowell County was over 90% rural, and Mercer and Raleigh counties were approximately 70% rural (West Virginia Bureau of Employment Programmes, 1993).

Industrialisation in Southern West Virginia: Expansion and Technological Change in the Coal Industry

The area's integration into a national industrial economy involved a steady expansion of labour intensive coal mining operations between 1880 and 1930. High-quality coal, especially the 1.8-3.7 m thick Pocahontas Number 3 seam, outcropped in the south-eastern end of McDowell County, and drew outside interest as development began in the early 1880s. By that time, the Pocahontas Fuel Company mines were producing near Pocahontas, Virginia and by the turn of the century, operations had diffused across McDowell County and to massive works at Itmann in Wyoming County (Pocahontas Fuel Co., 1936). The resource base in the two counties was vast; the original winnable coal reserves amounted to over ten billion short tons (West Virginia Chamber of Commerce, 1987). The coal met a broad range of markets, especially metallurgical uses, and as home heating and thermal-electric fuels. These qualities enjoyed a valued reputation in a market area that reached from the eastern Seaboard to the Midwest and Great Lakes (Murphy, 1933; Gillenwater, 1977).

Firms such as the Pocahontas Fuel Company and US Steel acquired land and mineral rights, built towns and invested heavily in mining infrastructure. They controlled the majority of the coalfield's production and marketing, operating 'captive mines' which served their firms exclusively. Individual firms took an enduring prominence in resource development decisions. As late as the 1980s, for example, over 86% of McDowell County's mineral area was held by major steel and energy corporations (Appalachian Land Ownership Task Force, 1983).

Coal mining technology at the turn of the century relied on labour-intensive methods as coal was hand cut and loaded in underground development processes. Machine cutting was widely adopted by 1930, but 90% of the coal mined continued to be loaded by hand. As a result, a labour force of perhaps 500 to 1,000 miners was required to run coal from a single mine (Munn, 1977; Yarrow, 1979). In the regionally formative period between 1880 and 1930, approximately 110 mining towns were built in support of 225 mines developed in southern West Virginia (Gillenwater, 1977). Employment mushroomed during and after the First World War, increasing housing demands that largely were met by mine operators. By the mid-1920s, West Virginia held half of the nation's company mining towns (Munn, 1977). The practice became pervasive over time and by the early 1930s, just over four of every five miners in West Virginia lived in company-owned houses (Magnusson, 1932). In that environment, coal operators commanded considerable social and economic control. Near monopolistic advantages in jobs,

125

housing and wages were in their hands as were company stores which frequently dominated retail trade in the towns (Corbin, 1981).

More than 50 mining firms had partitioned coal lands into leased tracts along the creeks of McDowell and Wyoming counties. The most extensive tracts were held by the earliest developers, the Pocahontas Fuel Company and US Steel's subsidiary, the United States Coal and Coke Company. Pocahontas' operations spanned much of the study area and included over 35 mines. US Steel entered the field a bit later, but had developed seven different mines around its Gary works (Pocahontas Fuel Co., 1936). Other leases were acquired by firms which were subsidiaries of other iron and steel makers in a widespread development of metallurgical quality coal.

As the centre of the most active development, McDowell County's economic structure was dominated by the coal industry in 1930. Its 19,000 miners accounted for 65% of the county's total employment. Mining in Wyoming County developed somewhat later, but it also relied heavily on the coal industry. A mining labour force of 1,940 comprised 31% of that county's total employment. In each case, the importance of mining employment is probably understated because spin-offs from the single-industry development touched virtually every other sector of the economy (Moore, 1998). The area's mining economy produced a remarkably specialised economic landscape (Murphy, 1933). That landscape symbolised the area's connection with and dependence on the national industrial economy.

The Appalachian coal industry experienced substantial technological restructuring following World War II. Mechanised cutting and loading machines transformed the once labour intensive industry, displacing thousands of miners. Between 1940 and 1963, national bituminous coal employment fell from 440,000 to 142,000, without a parallel decrease in the quantity mined. For West Virginia and other Appalachian coal producing areas, these developments meant nationally high rates of unemployment and poverty (Estall, 1968; Raitz and Ulack, 1984; Couto, 1994). In a telling indicator of changing economic conditions, eastern Kentucky, West Virginia and south-western Virginia lost nearly two million of their population to outmigration between 1940 and 1960 (Raitz and Ulack, 1984).

Southern West Virginia, on the strength of its metallurgical coal markets, initially was insulated from the larger structural adjustment in the coal industry. Employment in 1950 had risen above the 1930 level, solidifying the position of the industry in the local economic structure. For example, mining employment in McDowell County continued to comprise two-thirds of the total employment and in Wyoming County the industry had grown to account for half the county's jobs (Moore, 1998). Mechanised loading, however, virtually replaced hand loading after 1950 and machines increasingly were used to undercut coal seams. By 1970, the productivity per miner per day in West Virginia's underground mines reached 12 t, a rate of output that had tripled since the turn of the century. Even more efficient technologies were realised with the improvement of long-wall mining techniques and continuous mining equipment. Under good conditions, a typical continuous mining unit worked by a crew of only 8 men could produce 350 to 400 t in a shift (Yarrow, 1979). West Virginia's mining employment ultimately was sensitive to these changes. It peaked at almost 126,000 in 1948 and by 1970, had declined to just over 48,000 (Zimolzak, 1977). The region's continued dependence on a restructured mining economy made the area even more vulnerable to future shifts in the national economy, and especially to those which occurred within its traditional industrial linkages.

Deindustrialisation and Post-industrialisation of the National Economy

Service sector employment accounted for virtually all of the net job growth generated by the American economy during the 1970s and 1980s. The shift away from industrially based economic growth was profound in a number of respects. First, it ended a century-long reliance on manufacturing as the primary engine of national economic growth and severely threatened America's status in globally-competitive manufacturing markets. Second, it held serious domestic economic and social costs. The capacity of the service sector to absorb displaced manufacturing workers was problematic. Furthermore, wages, benefits, and full-time employment opportunities in the service sector were far lower than in manufacturing (Rodwin, 1989). Finally, industrial restructuring operated across a number of geographical scales as plant closings and manufacturing job losses had tremendous impacts on traditional local and regional economic bases. It was a process that reshaped the regional geography of America's economic landscape (Clark, 1985; Markusen and Carlson, 1989).

Explanations of the restructuring of regional economies centre on a debate between those who see deindustrialisation as a corporate decision-making function aimed at disinvestment in conventional business linkages and those who see the process as a logical progression of economic maturity within industrially based economies. The latter, or post-industrial perspective, contends that in a dynamic economy, some industries and regions grow and decline in response to changing technologies and demands. Therefore, cyclical change and regional sectoral shifts are expected from influences inherent in a free market system. As those dynamics occur in mature, industrially developed economies, a 'hollowing out' of blue collar manufacturing sectors would be anticipated. The related force exerted by product cycles may explain the 'hollowing out' process. High-tech products of previous manufacturing eras, such as steel and automobiles, have become low-technology products compared with consumer electronics, semi-conductor chips and robotic machine tools. In the face of that product shift, manufacturing employment and production moved to high-wage rivals in Japan and Europe, undermining America's competitive advantage with more efficient production technologies and investment (Rodwin, 1989). Deindustrialists, concerned with the speed and direction of these trends, see disinvestment as a driving force in the decline of manufacturing's traditional prominence.

Regardless of whether one attributes employment losses in manufacturing to either a deindustrialist or a post-industrialist perspective, that sector's subordinate role in national economic restructuring represented a fundamental shift in the American economy. The American steel industry was particularly sensitive to the changing economic environment. In 1988, there were 169,000 steelworkers in the United States, down by two-thirds since 1958. Individual steel-producing states lost jobs at rates that varied from 40 to 60% over the three decades. In perhaps the

127

largest single-firm layoff, US Steel closed 14 mills and plants in eight states at the end of 1979 (Bluestone and Harrison, 1982). Between 1982 and 1985, steel centres such as Chicago, Birmingham, Youngstown and Pittsburgh lost basic steel jobs at rates from just under 30 to over 50%. Perhaps one-third of the losses were attributed to improved technology and automation; proportionately deeper cuts occurred as imported steel made increasing inroads into the American market during the 1980s (D'Costa, 1993). The reduction in domestic production was especially symbolic of industrial decline because the geography of the steel industry formed the core of much of the nation's industrial pattern. Impacts on local services dependent on wages from the steel industry were tremendous. Bluestone and Harrison (1982), citing the ripple effects of the 4,100 jobs lost in US Steel's 1979 closing of a Youngstown, Ohio mill, suggested that plant closings caused entire regions to behave like company towns. By comparison, Southern West Virginia was a region comprising company towns oriented almost completely to markets in the steel industry. The effects there were profound. They truncated economic connections which had spanned a century.

Deindustrialisation and Restructuring in Southern West Virginia

Deavers (1991) identified the 1980s as a decade of broad stress for rural America. Employment losses occurred in all sectors of the rural economy as jobs disappeared in agriculture, manufacturing and mining. For southern West Virginia's coal miners, the 1980s were a period of extreme stress brought on by a catastrophic contraction in their industry. Although much was made of a hopeful improvement in the Appalachian economy during the energy crisis of the late 1970s (Appalachian Regional Commission, 1985), the recovery was short-lived. The region's coal industry soon suffered under the compounding stresses of the 1981-1983 recession and national economic restructuring in traditional markets as manufacturing declined in the Midwest (Markusen and Carlson, 1989). Particularly damaging was a decline in the demand for metallurgical coal as steel production slipped.

Mining sector job losses between 1980 and 1990 were greatest in McDowell County where over 75% of the labour force was displaced (see Table 10.1). Over 60% of Wyoming County's mining jobs disappeared; in Raleigh County layoffs reduced the 1980 mine employment by more than half. Fewer than 1,000 miners worked in Mercer County, but virtually all of them lost their jobs. Perhaps more important is the relative weight of these losses to the total employment decreases in the area. Employment in all the counties declined substantially; coal industry jobs lost in those places accounted for a remarkable share of the decrease, underscoring a deep and widespread dependence on that sector (see Table 10.1). Roughly 8 out of every 10 employees who lost jobs in McDowell and Wyoming counties worked in mining. Unemployment rates in the two counties peaked at 30 to 40% in 1983 and at 20 to 30% in 1987. Through most of the 1980s, annual rates hovered around 20% (West Virginia Bureau of Employment Programmes, 1993).

Employment changes in McDowell and Wyoming counties reflected regional coal industry trends, especially those that concentrated production in the control of a few large firms. Over 50% of the production in the two counties was controlled by the two largest operators, US Steel and the Consolidation Coal Company (Zimolzak, 1977). In their operating decisions, two of the area's largest coal preparation plants ceased operations in the 1980s. The massive Alpheus Plant, operated by US Steel near Gary, in McDowell County closed in 1982, eliminating approximately 1,200 jobs at that site (*Mullens Advocate*, 1990). The following summer, all 7 of the mines near Gary closed; although all but one reopened a year later. Hopes for a return to past employment opportunities were dampened when the mines resumed operating with only a fraction of the previous labour force. Even in the depressed economic climate, rumours of re-hirings spread rapidly but proved to be untrue (Appalachian Regional Commission, 1984). After the re-opening in the summer of 1984, the mines shut down for good in 1986. The closing related directly to deindustrialisation in the steel industry. The mines' markets were US Steel's mills in Gary, Indiana and Clairton, Pennsylvania. US Steel then leased some of its coal seams to small operators who employed a total of fewer than 200 miners (*Mullens Advocate*, 1990).

Table 10.1. West Virginia total employment and mining employment change, 1980-1990

County	1980 mining employment	1990 mining employment	Absolute change	% change
McDowell	7,200	1,600	-5,540	-76.9
Mercer	890	60	-830	-93.3
Raleigh	4,860	2,270	-2,590	-53.3
Wyoming	4.610	1,700	-2,910	-63.1
		1990 total employment	Absolute change 1980-1990	Mining employment losses as % of total change
McDowell		6,720	-9,640	79.8
Mercer		25,530	-2,650	31.3
Raleigh		26,680	-4,490	57.7
Wyoming		5,790	-3,320	87.7

Source: West Virginia Bureau of Employment Programmes, 1993

At Itmann, in Wyoming County, the Consolidation Coal Company suspended operations in December 1986 at a preparation plant which had employed 1,350 at the peak of its activity. The Alpheus and Itmann preparation plants have been dismantled in mine reclamation efforts, erasing landscape records of a declining economic system and removing pieces of infrastructure that are no longer needed. Their absence on the landscape is a tacit statement of the end of a regional economic era. It also serves as an ironic footnote to the findings of a 1982 survey of McDowell County residents which revealed that nearly a third of the respondents ranked mining employment as their first choice among specific economic growth alternatives (Trent *et al.*, 1985). The McDowell County residents' response reflected an economic reality faced by many who remained hopeful of gaining a $36,000 average annual coal mining wage rather than settling for a $12,000 service sector wage (West Virginia Bureau of Employment Programmes, 1993). Simply put, service sector earnings would have to triple to replace lost wages in mining employment.

Even during its brief rebound during the late 1970s, the coal industry's renewal did relatively little to stimulate growth in other sectors of the Appalachian economy, especially in the service sector (Miller, 1978; White, 1989). The structurally weak economy of southern West Virginia included few catalysts for service sector opportunities. Changes in service producing employment between 1980 and 1990 resulted in only modest gains (see Table 10.2). In less rural and more accessible Mercer and Raleigh counties, the service sector employment rose by less than 10%. In adjacent Wyoming County, a slightly negative rate was recorded. This trend perhaps was moderated by the fact that 30% of the labour force commuted to jobs in nearby counties. Less than 10% of McDowell County's workers commuted to work places outside the county and service sector employment had been reduced by more than 1,100 jobs (West Virginia Bureau of Employment Programmes, 1993). The county's largest private sector employer was a single KMart department store which accounted for 100 jobs (Baldwin, 1995). A decreasing local population also limited the potential for expansion. From 1980 to 1988, McDowell County's net migration losses amounted to 11,200 and Wyoming County lost 4,800 of its population through migration (US Department of Commerce, 1991). With these local limitations, the most likely commuter destinations for service occupations were Mercer and Raleigh counties, where labour surpluses created keen competition in a new economy. The north-eastern and Midwestern industrial areas served for more than a century by these counties' resources recovered by restructuring towards a service sector economy. That potential did not exist in southern West Virginia and a similar restructuring did not occur.

Table 10.2. Southern West Virginia employment change: service producing sector, 1980-1990

County	1980	1990	Absolute change	% Change
McDowell	6,036	4,910	-1,120	-18.6
Mercer	19,120	20,120	+1,000	+5.2
Raleigh	18,530	19,950	+1,420	+7.6
Wyoming	3,870	3,850	-20	-0.5

Source: West Virginia Bureau of Employment Programmes, 1993

In 1998, McDowell, Raleigh, and Wyoming counties continued to be identified as economically distressed by the federal government. The classification was scaled on weighted averages of unemployment and poverty rates and low income levels which exceeded national rates by as much as 200% (Appalachian Regional Commission, 1998). Perhaps the most hopeful event occurred when McDowell County won a million dollar grant and federal designation as an Empowerment Zone (Baldwin, 1995). The county's success in acquiring the grant and the programme's emphasis on local involvement and entrepreneurship are positive signs that are sorely needed.

Post-industrial Southern West Virginia: A Trio of Third World Developments

The collapse of the coal industry and the subsequent lack of restructuring left the area very limited economic development options. High rates of poverty and unemployment within a stagnant economy made development issues even more pressing. Three developments mirrored the region's history of intervention from the outside and symbolised the fragile conditions of the late 20th century. In some instances, these relegated the area to a status that resembled dependent third world areas.

Landfills for Out-of-State Waste

Southern West Virginia was economically depressed, had large tracts of undeveloped land, had weak state regulatory agencies and was relatively close to large urban places in the North-East. All of these characteristics made the area a target for proposed landfill sites that would dispose of out-of-state waste. National waste management businesses from Chicago, the North-East, and Alabama were also attracted to the area's low disposal charges. In northern New Jersey, for example, the cheapest landfill charges were $110/t. In West Virginia, these charges ranged from $1/t to $20/t and averaged just $8 (DeHart, 1990).

By the late 1980s, Summers, Fayette and Raleigh counties were entertaining proposals that triggered a strong grass roots citizens' protest. Those contesting the deals recognised that many of the proposed landfill sites occupied abandoned coal mine areas. None of these were environmentally sound and some were not located far away from small settlements. The opposition was not grounded in the NIMBY philosophy; the projects posed real threats to safe water supplies and were at odds with investments in tourism, such as white-water rafting, that depended on environmental quality. The issue was made more complex by the revenues that disposal would generate for the individual counties (DeHart, 1990). Citizen protest was powerful enough to influence legislation that limited the size of landfills and generated more careful scrutiny in the permitting process. Landfills established in those counties have not been opened to out-of-state waste.

McDowell County did not possess the same proximity to the interstate highway system as the counties mentioned above. As the hardest hit by the economic downturn of the 1980s, though, it too became a prime candidate for out-of-state waste dumping. As early as 1987, citizens of the county stated that they needed jobs but did not want other people's waste. Over the next three years these sentiments took on a broader meaning as protests mounted against a Philadelphia firm's promotion of a 2,430 ha landfill site located on land long held by an absentee coal developer. The issue evolved into a heated local debate and became politicised at the state level. Plans for the project were accepted at first by the people of McDowell County and supported by Governor Gaston Caperton. In the face of activists' opposition, endorsement from the state was withdrawn, but a 1992 local referendum narrowly approved the project and was hailed as a triumph of local choice (Aquino, 1993). The results of the referendum were so close, however, that the debate continued and a consensus was not reached. The site, scheduled to open in 1995, did not. Today, residents who opposed the project proudly point out that their voice was heard.

Outside Relief Efforts

Religiously-affiliated efforts aimed at improving the quality of life in Appalachia date to the extra-regional recognition of the area's lagging development in the 1870s. The earliest attempts, for example, established 'Mission Schools' in eastern Kentucky and Tennessee. These educationally oriented programmes built schools and implemented a curriculum which taught basic skills such as agriculture and home economics. Perhaps because the perceived need was greater in the highlands to the west, the movement did not extend into Virginia and West Virginia (Strickland, 1993). Perceptions of eastern Kentucky and Tennessee as backward areas were influenced by a powerful popular media imagery in the late 19th century (Moore, 1991).

In the wake of the collapse of the coal economy a century later, McDowell County and Gary became campaign centres for on-going welfare efforts based outside West Virginia. Particularly active are individuals and a network of churches in the Charlotte, North Carolina metropolitan area (Garfield, 1994). Sponsoring agencies include a number of denominations. Baptist, Methodist and Presbyterian churches all carry out relief efforts and some have provided thousands of dollars worth of funding for ministries established in the region. Shipments of blankets and winter clothing leave for the area in the autumn, followed by boxes of toys at Christmas. During the summer, activities shift to housing rehabilitation and new construction in the remote hollows and depressed mining towns. These efforts are easily identifiable in messages on church marquees and in bulletins posted in local newspapers. Churches' message boards admonish passers by to 'Pray For Our Mission Trip to West Virginia', or to 'Donate Articles: West Virginia Mountain Mission Drive Underway.' Such volunteerism fills material and spiritual needs unmet by federal programmes and is reminiscent of similar programmes aimed at impoverished Third World populations.

The Charlotte-southern West Virginia connection is almost certainly founded in the reversal of a traditional Appalachian migration stream that was truncated by the loss of mining jobs and exacerbated by the disappearance of manufacturing jobs, which accompanied restructuring in the North-East and Midwest. The new southern stream began in the late 1980s and followed Interstate 77, a route that made Charlotte a destination accessible in just 3.5 hours from West Virginia (Martin, 1987).

Third World Poverty Tours

Regardless of its intentions, a San Francisco-based travel agency drew the ire of West Virginia state officials when it included places in the southern coal fields on its 'Third World in America' tour in 1990 (*The Herald Dispatch*, 1990). The programme in many ways serves as a metaphor for the recent development of the area. That it evoked strident local reaction attests to the lingering stresses of economic change in the 1980s.

Conclusions

Not all rural systems are as intimately connected to the impacts of national economic restructuring and deindustrialisation, especially at the geographical scale and conditions unique to the Appalachian region. Nevertheless, the larger issues of economic change present adaptations that must be met by rural places with limited economic bases. In this context, the Appalachian experience demonstrates the deep effects of national economic restructuring on rural economies and holds important insights into broader rural development venues.

Southern West Virginia's poorly developed service sector was fundamentally unlike those of urban places which took on unusually productive roles in economic restructuring. Quite simply, the regional service sector could not function as an economic engine to redirect growth. That lack of development resulted from a century-long reliance on an economy comprising a single industry. Furthermore, the economic system was dependent on a robust performance of a larger business climate operating outside of and in control of the smaller regional economic environment. This circumstance represents a difference in degree, rather than in kind, to those operating in single industry textile mill towns or in rural places that rely on the performance of branches or subsidiaries of urban-based manufacturing firms.

For other resource-based economies, the management and development of Appalachia's resources is a testament to the need for foresight and the uncertainty that unchecked exploitation and changing technologies may bring to sustained development. The potential for low-wage service sector employment to replace jobs lost in restructuring raises serious doubts, particularly with regard to providing income levels that translate to quality of life issues. For Appalachia, and other rural places these are lingering, problematic policy issues. Finally, the emergence of Third World style economic alternatives and perceptions is troublesome. In an era where imagery and reality are separated by only blurred lines, negative regional labels often create developmental barriers that are difficult to overcome.

Chapter 11

'Enabling' Technology for Disabled People: Telecommunications for Community Care in Rural Britain

Robert Gant

Disabled People in Rural Britain

The development of post-modern and post-structuralist ideas in rural social geography has been well documented (Cloke *et al.*, 1994; Phillips, 1998). Traditional, elitist and ordered gazes on rurality have been seriously questioned and issues related to identity, exclusion and power relations now form a substantive part of a re-focused research agenda (Milbourne, 1997). This reorientation towards social difference and 'otherness' embraces interpretative studies of the circumstances and experiences of marginalised people (Philo, 1992; Sibley, 1995). Debate continues, however, on theoretical perspectives for studying the geography of disabled people, including those who experience mobility-impairment (Golledge, 1996; Imrie, 1996; Gleeson, 1996). Ableism, which assumes able-bodied status, has been challenged intellectually and in political practice (Chouinard, 1997). New and critical geographies of disabling difference are beginning to emerge framed within concepts of domination, enabling justice and spaces of resistance (Gleeson, 1997). The research focus has consequently shifted beyond geographies which make visible the lives of disabled people to critical conceptions and explanations of disablement as a socio-spatial process in which physical or mental impairment become the basis of disempowerment and oppression (Imrie, 1997; Parr 1998). As yet, however, few studies have specifically engaged with the social exclusion and related life-styles of elderly and disabled people in rural environments (Buchanan, 1983; Gant and Smith, 1991; Gant, 1995).

Within the European Union this critical issue of ageing and related mobility-impairment demands wider consideration given the policy consequences of budgetary constraints and regional diversity in demographic trends (Commission of the European Communities, 1995). It is forecast that in the year 2020 25% of the population will be aged over 60: the largest increase is expected in those aged over 75 for whom disability is most prevalent (Knipscheer, 1994). Half the total population already lives in rural regions where the quality of life is threatened and problems of economic re-structuring have to be addressed (Commission of the European Communities, 1991a). In Britain age-related disability presents an equally significant challenge to the delivery of statutory welfare services. Of the total 6.2 million disabled people identified by an OPCS survey in 1988, 4.3 million experienced some difficulty with locomotion (OPCS, 1989). Further restrictions are being imposed on personal mobility as chronic mental and physical disability are compressed into a shorter and later phase in life (The Alzheimer's Disease Society, 1992). The Commission of the European Communities has recognised the important role that customised telecommunications can play in sustaining an independent lifestyle for disabled people. In 1991 the TIDE (Telematics for the Integration of Disabled and Elderly people) research programme was launched with the objective of reducing personal dependence, enriching social networks and containing the immediate cost of care (Fozard *et al.,* 1994).

Notwithstanding the opportunities presented by various forms of 'assistive technology', it must be acknowledged that personal disability and related social exclusion are both experienced and contested in a rural context. This is strikingly illustrated in the British countryside where, since 1945, far-reaching changes in the geography of rural transport and communications have compounded and differentiated the mobility experiences of elderly and disabled people (Nutley, 1988; Cassidy, 1994; Gant, 1997). Irrespective of the provisions of the Disability Discrimination Act 1995, many people still remain disadvantaged with respect to their travel needs (Frye, 1997). A re-evaluation of telecommunication services for mobility-impaired people is long overdue. Although Tinker's seminal study *Telecommunication Needs of Disabled and Elderly People* (1989) provides a qualitative assessment of personal needs and levels of satisfaction, it lacks a differentiated rural focus. The Rural Development Commission/OFTEL study *Telecommunications in Rural England* (RDC/OFTEL, 1989) has similar shortcomings. In contrast the few published studies on the adoption of personal alarm systems (McGarry, 1985; Fisk, 1989) and use of telephone services (Wenger, 1990) provide revealing insights into the coping strategies of vulnerable groups in contrasting rural environments.

Local authorities in the United Kingdom were required to publish formal plans for community care under the provisions of The National Health Service and Community Care Act, 1990 (HMSO, 1992). These three-year plans attach particular importance to the needs and circumstances of elderly people, especially those who are dependent and/or frail, people with a physical or sensory disability, and those with mental handicaps or learning difficulties. The aims of policy are clear: to permit independent living and provide support in coping effectively with changes in personal needs and circumstances (Henwood, 1992). These concerns are consistent with the wider objectives of the telecommunications industry to provide 'telepresence' as a service, not as an isolated set of technologies. Walker and Sheppard (1997, p. 14) articulate the case: "The starting point for all communications is a customer need, a particular desire for contact. The nature of that need is then reconciled with the available technology in selecting an appropriate service. At present, the choice of technology for many people may be restricted to a phone call, fax, post or a physical meeting, with e-mail an increasingly prevalent option. However, we can anticipate a future of greatly increased diversity of media, where there will be a much wider choice in matching an appropriate degree of presence with that initial desire for contact". Within the context of community care and wider issues of mobility-impairment, this study evaluates existing home- and community-based telecommunications services in the North Cotswolds (England) and argues the case for the development of a 'low-cost-voluntary-care-support-model' responsive to client need, mobility status and community setting.

Community Care in the Cotswolds Environment

Three policies which were crucial to daily life in the North Cotswolds were identified in the *Gloucestershire Community Care Plan 1991-1994*: firstly, to work with the strengths of the community, to recruit volunteers, and to provide mutual help and support amongst elderly people so that overall provision is improved; secondly, to enable the development of transport services for elderly people so that their isolation is reduced and they are able to make best use of available services; thirdly, to enable the expansion of domiciliary care services to meet individual need in all areas of the county throughout the week, day and night, and offer a realistic alternative to residential care (Gloucestershire County Council, 1991). The cyclical preparation of personal care plans within this framework raises serious questions regarding the age profile, mobility status and lifestyle of clients (Caldock and Wenger, 1992).

For the constituent groups of elderly and disabled people, the processes of rural change impose a particularly challenging set of problems. In 1991 57% of the 63 parishes in the North Cotswolds had populations below 350 (Fig. 11.1). Bourton-on-the Water (population 2,907) was the largest market town; Northleach (population 1,460) the smallest. Since 1945 the popularity of the region for retirement and the out-migration of younger elements have progressively distorted its age structure. An increase of 39% in the pensioner population in the intercensal period 1971-1991 reflects this trend. In 1991 27% of the 29,000 residents were pensioners, and 10% were aged over 75. Just over one third lived in the five market towns; the others were scattered widely in the main villages and rural parishes. From a welfare perspective, it is significant that lone-pensioners accounted for 19% of households; of these 51% lived in scattered villages and isolated hamlets.

These demographic changes have coincided with a reduction in the level of community services and their distribution in the settlement hierarchy. The North Cotswolds lacks a significant urban focus: for higher-order health and shopping services journeys are made to towns outside the region. At present the five small market towns support a balanced, if minimal, selection of health care, professional and retail services. The main villages, in contrast, normally support an Anglican church and a few low-order services, including a post office-cum-shop. Post offices, however, are being threatened by commercial procedures for

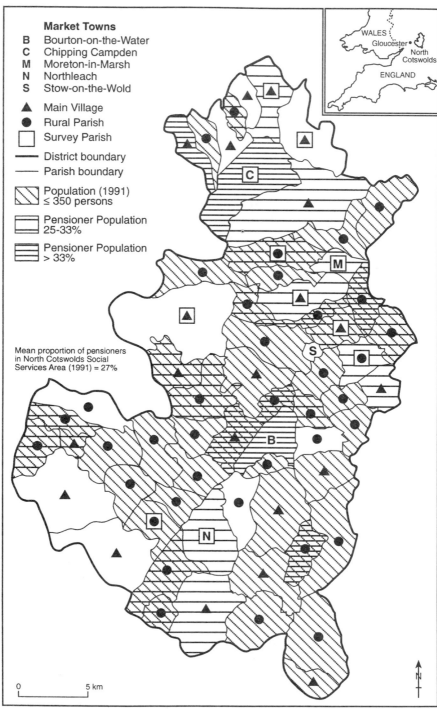

Fig. 11.1. North Cotswolds - Distribution of Pensioners and Population 1991
(source: Census of Population Small Area Statistics 1991)

rationalisation and village shops by declining threshold populations. Where they operate mobile vendors of groceries provide a lifeline for the elderly living in rural parishes which have lost their complement of basic services. The coincident decline in public transport services across the region has aggravated this situation.

Survey Design and Mobility-impairment

During the Autumn of 1993 quota samples of 286 households with elderly and disabled members were interviewed in 11 representative parishes in the North Cotswolds. These parishes were selected from a cluster analysis using Ward's method applied to 23 census and disability variables reported in *North Cotswolds Surveys: Parish Profiles* (Gloucestershire County Council, 1986). The pre-tested questionnaire covered details of: household structure; characteristics of accommodation; health and mobility status of members; life satisfaction; the support provided by friends, neighbours and relatives; contacts with community and statutory services; personal mobility and access to transport; typical journey patterns; and the design and use of personal alarms, telephones and other telematic services. These baseline data were essential for the critical evaluation of existing telecommunication services and development of customised systems for mobility-disadvantaged people in the rural environment (Gant and Walford, 1998).

For the analysis of personal mobility, the 458 people enumerated in the 1993 survey were divided into two groups: the elderly-mobile (361) who reported no difficulty in satisfying their daily needs for movement inside the home or for local journeys outside; and the mobility-impaired (97) who could not leave home and registered varying degrees of difficulty in moving indoors and in satisfying self-care and home-care needs (Gant and Walford, 1998). These two groups of elderly people were evenly distributed throughout the North Cotswolds by type and size of settlement, and the age and tenure of accommodation. Members of both groups suffered from a comparable range of physical and sensory impairments, but were strongly differentiated with regard to severity and multiple incidence. This was especially so with the age-related occurrence of Alzheimer's disease, which affected some people already afflicted with arthritis, circulatory and sensory impairments. The mobility-impaired, however, were characteristically older and proportionately more were registered as disabled (44%) and lived alone (43%). They also relied more heavily on special aids, particularly a walking stick/frame (59%) or a wheelchair (18%).

Group differences in mobility were reflected in levels of car ownership, travel patterns using a local service bus and typical journey destinations. Overall, 71% of the mobile group owned either a conventional vehicle or one adapted to compensate for infirmity; the corresponding proportion for the mobility-disadvantaged was 37%. This disparity in personal access to transport mode is reflected in travel patterns. Relatively few members of either mobility group used public transport and most of the car-less and more severely disabled received periodic visits from a GP and the chiropodist. In the larger settlements most of the people who were able to leave their homes walked to locally available facilities, whereas in the main villages and rural parishes which lacked even basic services they were forced to rely on a household car or lift with relatives, friends or neighbours to a local market town or larger regional centre.

Serious welfare problems can accrue from mobility-impairment and transport deprivation. Throughout the region, however, informal community care networks played an important role in supporting the lifestyles of elderly and disabled people. Much-appreciated friendship and neighbour networks existed in the market towns and surrounding villages. Informal, but discreet, surveillance was commonly practised; this was reinforced by daily contacts and reciprocal home visits for almost half of those who need help. Family links, too, were important. Almost equivalent proportions of the mobile and mobility-impaired (59%) had close relatives living in the same settlement. Those without close family support had to depend on informal care assistance provided in all localities by friends, neighbours and (paid) helpers; others had to rely more heavily on statutory social services comprising a home-help, district nurse or health visitor. Some, inevitably, had to cope alone.

Telecommunications Solutions: Reality and Prospect

Given the mobility profiles of elderly people in the North Cotswolds, the issue of accessibility to essential services needs to be reformulated to include an evaluation of new and enabling technologies based on conventional telephony and more advanced telematic applications. In a rural environment the household telephone is an important lifeline; it also provides a platform for telematic enhancement (Wenger, 1990). It can readily function as a substitute for personal travel. In 1993 90% of those interviewed had at least one household telephone. Both mobility groups had broadly similar patterns of telephone contact. Only 12% regarded the telephone as a luxury; for around half, however, the quarterly rental charges represented a significant burden on household budget. Understandably, those who experienced difficulties with personal mobility derived a marginally greater benefit from a mobile phone and equipment modified to suit sensory and reaching/handling impairments (British Telecom, 1994). Around one-fifth of each group had already been advised to seek modifications to existing apparatus to meet their personal needs, but had failed to do so. Relatively few elderly households, however, owned a personal computer.

A telephone meets the dual need for emergency contact and informal personal support. Around 70% of those interviewed considered a telephone essential for contacting a doctor, family and friends; in contrast, the proportions declined to 20-30% for contact with shopkeepers, the Social Services and a Community Nurse. Overall, 34% of those interviewed claimed to have used their telephone in an emergency during the past year. For those who lived alone it was the most important means for summoning help in times of dire emergency. Two low cost user-friendly schemes which meet the requirements for contact and emergency help already operate successfully in parts of the region: the 'telephone tree' and Cotswold Careline Alarm System. These are based on the household telephone and demand vigorous promotion to benefit those most in need, especially in the rural parishes.

Telephone Tree

Variants of the 'telephone tree' exist in several Cotswold parishes. The 'tree' is formally structured and operates as a 'cascade'. Each member has a fixed place in a network and, having received a telephone message, initiates contact with two (or more) others. Triggered by events such as severe winter-time blizzards, influenza epidemics and the need to disseminate information concerning community affairs, this relatively low-cost system can be readily extended through verbal contact to include those without telephones. It overlays the loosely-structured support networks commonly identified in the survey and reinforces patterns of face-to-face contact, family visiting, and informal neighbour surveillance. It can respond quickly to the changing personal needs for support within the community and is relatively easy to organise. In several Cotswolds parishes local Care Committees spawned by Church organisations and parish councils are actively seeking to extend such formalised networks for telephone contact. These networks have the distinct advantage of becoming meshed into statutory care provision and can effectively capitalise on high levels of telephone ownership in thinly-populated parishes where immediate and more casual surveillance is less effective.

Cotswold Careline Alarm System

Cotswolds Careline comprises a 24 hour-a-day computerised control and response system administered by the Cotswold District Housing Department via Gloucestershire Ambulance Service. In 1992 it had 1,400 linked members with telephone and 'pendant' alarms, some of whom live in sheltered accommodation. During a sample period 1 October to 31 March 1992, 5,227 calls were logged. The majority was raised by user error, routine checks and administration. Only 299 (6%) were genuine requests for assistance; two-thirds were directed to wardens in sheltered accommodation. In July 1992 the most favourable ratio of alarms to eligible persons was found in the market centres and villages which had active Care Committees. Voluntary agencies, health professionals and family members are active in promoting the adoption of personal alarms. One third of the households with mobility-impaired people and one sixth of the remainder had a member who relied on a personal alarm. In general those with alarms found them easy to use and appreciated the reassurance. One fifth of lone-residents with mobility impairments claimed that they would use their alarm in time of emergency to summon help, a proportion almost equivalent to those citing the telephone. Those without personal alarms were normally aware of the Careline scheme, its operation and immediate benefits, and knew of someone provided with an alarm. Notwithstanding advice, and even personal demonstration, some were not yet prepared to commit themselves to adoption.

Future developments in care provision can effectively build on the widespread ownership of a household telephone. Rapid technological change is a characteristic of the telecommunications industry. No longer is it appropriate to discount the widespread development of Community Teleservice Centres (CTCs) or electronic mail systems and cable services; and links with the Internet have become a reality for even the most isolated homes in rural Britain. For access to more advanced and powerful telematic and community-based telecommunications services, the options provided by CTCs and electronic mail networks need to be evaluated with regard to the communication needs and existing patterns of community support for mobility-disadvantaged groups in the Cotswolds environment.

Community Teleservice Centres (CTCs)

Gloucestershire Rural Community Council has been actively engaged with other agencies in the county in promoting the development of a network of Community Teleservice Centres (CTCs) or telecottages. This early initiative, the Gloucestershire Telecentre Network, has temporarily foundered, however, as a consequence of the withdrawal of countywide commercial sponsorship (Harris, 1995). Notwithstanding this short-term setback, it is timely to make a realistic assessment of the social potential of CTCs with regard to care provision for mobility-impaired people within the scattered distribution of small market towns and villages in rural Gloucestershire. CTCs are well-established in Scandinavian countries and by July 1996 over 150 had been established in rural Britain (Denbigh, 1996). Their main role is to provide people in rural areas with access to data-processing equipment, telecommunication facilities and computer-assisted information services. Individual households are not immediately tied to the network; a journey is required from home to use the cluster of services available in a parish hall, schoolroom or other accessible, but secure, building.

Almost two-thirds of the North Cotswolds parishes, however, have populations of less than 350. Given such low thresholds of demand, it seems unlikely that the Scandinavian model of the fixed CTC will be sufficiently flexible to satisfy the basic needs of the elderly and disabled for a range of appropriate information services, telecommunication linkages and social contacts. Furthermore, the *raison d'être* of the CTC is business. Without a sound financial base, or external subsidy, the welfare needs of local people can only be satisfied through payments at an economic rate. This will inevitably act as a deterrent to those on low incomes and possibly stymie the diffusion of telecommunication linkages within the local community. Whatever the form of the CTC, training and assistance for users would be an immediate requirement. In Cotswold market towns CTCs could be sited in library buildings or schools. For the more sparsely populated areas, variants of the CTC based on a mobile platform should be developed experimentally and evaluated in relation to care needs and population distribution. In this connection Gloucestershire County Mobile Library vans can provide an effective and regular service to the majority of parishes with less than 350 residents, including significant proportions of elderly and disabled people.

Electronic Mail Networks

During the 1990s the National Health Service in Britain has been progressively re-structured. Information Technology is being widely used to enable a more market-driven health sector and greater focus on community care (Bowles and Teale, 1994). Information management and technology are vital to achieve the objective of developing a person-based, integrated, secure and confidential system of health and community care. Electronic mail networks provide a first and primitive stage in this process: more sophisticated, screen-based, high speed data services have already been developed experimentally for telecommunication support (Richardson and Riley, 1994). More immediately, however, and as a first step there are opportunities in rural areas to introduce, develop and co-ordinate e-mail networks for the two-way dissemination of requests for information and advice.

The potential of this service has been defined, in an urban environment, by the Manchester HOST which was launched in 1991 as the first locally-controlled public communications and information system in the United Kingdom (Graham, 1991). Disabled people were among the several 'economically marginalised' groups targeted by 'HOST'. Through a network of Electronic Village Halls (EVHs) the prototype has demonstrated its potential for information diffusion via electronic notice boards, skills training and personal development. At the lowest level the e-mail network depends on the home-based facility of a computer and modem. However, the network can be readily extended to link up CTCs or other nodes where members of a community gather, for instance a village hall, school, post office or shop. The issue of funding remains a complex, but not insurmountable, problem. It needs to be addressed in an appropriate forum and as part of a comprehensive strategy for rural development (Grimes, 1992). This implies that the design of telecommunication services driven by local economic development should be sufficiently flexible to accommodate the communication needs of vulnerable, yet dispersed, groups like elderly and disabled people.

Progress Towards Telepresence

Field evaluation of existing provision, and consideration of experimental models in the Cotswolds points clearly to the need for low-cost, user-friendly telecommunication support which builds on known and trusted relationships within the community. Survey evidence shows that the telephone is the principal means of telecommunication. More advanced systems of telecommunication support have yet to be introduced. There is a heavy dependence on the telephone for emergency contact. Although it supports more casual forms of social interaction, the telephone is not a direct substitute for the caring support and warm conversation of co-residents, relatives, neighbours and friends in the ambient community. These relationships are unquestionably vital at advanced stages in the life cycle.

Figure 11.2 presents a conceptual model which combines these essential characteristics and responds realistically to individual circumstances, the

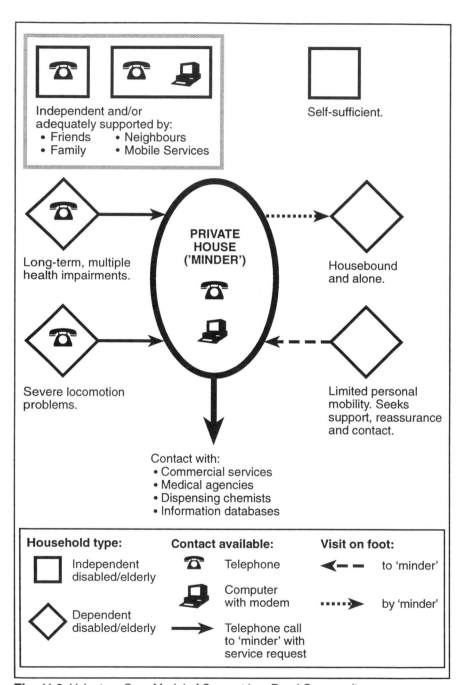

Fig. 11.2. Voluntary Care Model of Support in a Rural Community

importance of trusted personalities in local environments and levels of available service provision. It acknowledges the requirements of different fractions among elderly and disabled people in the countryside, their needs for support in home environments, and emphasises the importance of key and public-spirited individuals in the community who have as a minimum requirement a telephone with an attached modem. These volunteers assume the role of 'minders' for those individuals and households without a telephone or alternative network of social support. They also act as 'gateways' for households with basic telephones, but who need access to a wider range of computer-accessed information services and agencies.

In sparsely-populated areas, members of care committees drawn from neighbouring parishes might offer their support as 'minders'. Likewise, the provision of mobile telecommunication services based on the library van could be supported by a suitably trained attendant. Furthermore, in larger villages and the market towns CTCs housed in a primary school or village hall could provide an enhanced level of service both for visiting clients and nominated supporters who provide a caring link with housebound people. Of course, the cost implications for individual households and the County Social Services budget cannot be overlooked. Equally important, it must be recognised that innovation will impose an even greater responsibility for training and support on 'carers' and, in particular, the voluntary sector.

The promise of more advanced telematic services for the delivery of community care, however, should be carefully evaluated. The advent of broad-band and cable television with the capacity to carry multi-media and interactive services, heralds a new dimension of telecommunication support which will eventually benefit mobility-disadvantaged people in rural Britain (Greenop et al., 1994). As technical and service issues are resolved, an increasingly flexible and ubiquitous multi-media telepresence will progressively displace basic telephony in selected 'vertical markets' with clearly defined application needs and quantifiable benefits (Walker and Sheppard, 1997). The potential, for example, of wider-ranging telepresence and information services, teleshopping, telemedicine and screen-based teleconferencing for supporting new lifestyles in the countryside should not be dismissed from a present day mindset on grounds of use, cost and training needs. As Thornton (1993) argues, services should be so designed to sustain people in everyday living and not solely as triggers for emergency help. The time is ripe for further experimentation with community-based telematic services in the British countryside. Questions of sponsorship, management and evaluation need to be addressed, and actions agreed. The voluntary care model of support, however, presents a first and significant stage in linking telecommunications technology to changing personal needs in rural environments undergoing significant changes in their demographic structure and social composition.

Chapter 12

Production and Consumption in Rural Service Provision: The Case of the English Village Pub

Ian Bowler and John Everitt

Introduction

The geographical literature on the restructuring of rural services has tended to focus on features of either access, especially in the context of personal mobility and deprivation, or provision, usually in the context of the contraction and reorganisation of public services (Robinson 1990, pp. 344-363). In contrast, this paper explores a production-consumption approach to understanding the restructuring of rural services and for a private-sector rather than public-sector service, namely the English village public house (pub). In this context, 'production' concerns the physical fabric, the services offered and the product image of the village pub. 'Consumption' involves customer visits and the purchase of the various services provided, including the image of the village pub.

Lievesley and Warwick (1992) identify the village pub as one of the six 'key services' necessary for the creation of a more 'idyllic' rural life (Heap *et al.,* 1993), the others being a shop, post office, school, village hall/community centre and bus service. But like these other rural services, English village pubs are passing through a period of significant restructuring. Compared with the other five services, however, the village pub is under-researched (Clark and Woollett, 1990), despite its traditional function of providing a significant recreational and entertainment service for rural communities. Indeed the pub, described by Dr Samuel Johnson, the eighteenth century lexicographer, as "the throne of human felicity" (Allsop, 1993), plays an important role in the wider English socio-economic structure. Denter (1994), taking rural and urban pubs together, estimates that 29% of the adult population of Britain still visit one of the country's 61,000 pubs at least once a week; pubs and breweries provide employment for over 400,000 people (313,000 staff and 87,000 licensees); the average male adult drinks 102.6 litres of beer/lager annually; and the turnover of the pub trade is £20 billion a year. Monckton (1969) observes that the pub remains a significant national phenomenon and cultural symbol.

Four processes have been identified that connect production and consumption in the restructuring of the English village pub: (a) the investment policies of national and international brewing companies; (b) recent United Kingdom (UK) regulations for the brewing industry; (c) changing social behaviour of the new rural service class; and (d) changes in the leisure behaviour of the non-rural population. However, these processes of change in British brewing, pubs and drinking habits (Howkins, 1989) should be placed in the broader perspective of an international restructuring of eating and drinking services in general. For instance, Simons (1994) has identified the French bistro as a 'vanishing breed', with some 4,000 of these 'homey cafes' going out of business each year. Similarly, eating and drinking establishments in Anglo America, the acknowledged leader of popular culture change in the western world, are also constantly evolving (Crosbie, 1994).

Production of the Village Pub and Changes in the Organisational Structure of the UK Beer Market

Beer and lager sales still form a central if contracting part of the pub business and an understanding of the restructuring of village pubs must begin with an examination of their changing ownership structure and the relationship with industrial capital in the form of national and international brewing companies. The brewing industry has been changing throughout its existence (Dixon, 1978; Hawkins and Pass, 1979; Watts, 1987, 1991) but recently the pace and scale of change towards a few, large, multinational companies has accelerated. Indeed the brewing industry has been subjected to the same economic forces as other sectors of manufacturing industry, with an increased emphasis on global capital accumulation, the restructuring of production plant and employment, and the search for scale economies. The world market is now dominated by a few major brewers, each controlling a large number of well-known brand names (for example: Budweiser, Coors, Miller, Holsten and Kronenbourg). Although none of the top six are British, these brewers all market their produce in the UK.

The UK market is dominated by Scottish and Newcastle PLC (with John Courage - 28% of total sales) and the Bass Group (23%); the other major brewers are Allied Domecq and Whitbread. Together these brewers control 80% of the UK beer/lager market. On the other hand, the development of micro breweries serving local, niche markets provides evidence of the effects of more recent flexible business practices (Milne and Tufts, 1993); the major breweries have attempted to capitalise on this trend by reintroducing traditional beers under the brand names of companies taken over decades earlier. Consequently regionally named if not produced beers, and the advertising that accompanies them, are being re-established.

The structure of the companies is complex. The Bass Group, for example, with a sales turnover of £4.4m, is sub-divided into Bass Taverns (pubs), Bass

Breweries, Bass Hotels and Restaurants, and Bass Leisure (gaming machines, bingo clubs and bookmakers). Until recently the brewers also owned most of the retail outlets (tied pubs) for their products; tenants or managers were responsible for the pub businesses, and the independently owned and operated pubs (free houses) were in the minority. Under recent UK regulations (see below), the large brewers have been required to sell off many of their tied pubs to independent pub-owning companies. This changing pattern of pub ownership has been complex; for example, in the early 1990s Grand Metropolitan (Watney Mann) traded its breweries to John Courage (Elders/Fosters) in exchange for that company's pubs, thereby promoting John Courage to second place in the brewing 'league'; when in turn Scottish and Newcastle acquired the brewing interests of John Courage from the Fosters Brewing Group in 1995, the company became the market leader. During 1997, Bass showed an ultimately abortive interest in acquiring Carlsberg-Tetley from Allied Domecq, while in 1998 Grand Metropolitan and Guiness proposed a merger to form GMG Brands. Such developments would further increase concentration within the drinks industry, leading to another round of restructuring for breweries and pubs. By mid-1998, the main pub-owning companies had become the Grand Pub Company (owned by Nomura of Japan), Allied Domecq, Bass and Greenalls. While 44% of pubs are free of ties with a brewing company, the remainder are either managed for a large company (18%), run by a tenant/lessee for a national brewer (16%), or managed/tenanted for a regional brewer (22%).

The changed ownership of pubs has had two effects. Firstly, maximising economic profitability has been imposed on the pub as the major criterion for remaining in business. The new pub-owning companies have replaced traditional low-cost tenancies with high-cost 10 or 20 year leases. A lessee, unlike a tenant, does not have the protection of the 1992 Landlord and Tenants Act, has to pay a higher rent for the pub, is responsible for all renovations and repairs to the pub, and often has to meet sales targets laid down by the company or else pay a fine. As a result, of the 60,000 'houses' in the UK, 2000 have been closed since 1989 and a further 6000 are forecast to shut by the end of the century (Gates, 1993). Symptomatically, one of the three pubs in the case study village discussed below had closed within the last five years.

Secondly, new capital investment has been made by the brewing companies in their remaining national or regional 'pub chains'. Now each company has its own design and development department, each creating a catalogue of pub designs from which a selection can be made when investing in a pub development. In this way the brewing company's (or tenant's/lessee's) concepts of 'popular' image can be imposed on an increasing number of public houses, a trend found in the refurbishment and changed services of pubs both in rural and urban areas. In rural pubs it has resulted in the remodelling (conversion) of the internal structure of the pub to create large, standardised, open rooms, rather than the individualistic, sub-divided, small-room structure of the traditional pub with its public and lounge bars. Natural timbers and artefacts have been replaced with replica materials; furnishings have been remodelled to conform with contemporary urban rather than traditional rural standards of comfort and design; and spaces for restaurant and children-oriented services have been created.

When taken together, the penetration of wholly commercial considerations and the restructuring of ownership have served to reposition the village pub within the recreation and entertainment industry. The simple range of services traditionally produced for the village community have been replaced by building structures and decor designs that are at the same time uniform and homogenised, but also constantly in flux so as to reflect changes in the popular demand by an increasingly non-village clientele.

Production of the Village Pub and the Influence of UK Regulation of the Brewing Industry

Recent regulations have played a significant part in changing the business practices of the village pub (Everitt and Bowler, 1996). Firstly, stricter national drink-driving laws have reduced both the number of car-based eating/drinking trips into the countryside, as well as overall consumption of alcohol. Secondly, European Union fire safety and environmental health (hygiene) regulations are being applied to pubs. This has resulted in the need for additional capital investment in kitchen and toilet facilities and higher costs as regards staff training.

Thirdly, UK governmental intervention in the brewing industry has taken place, following from the 1989 Monopolies and Mergers Commission Report (MMC, 1989). Intervention by the Conservative government stemmed from the belief that the brewers, and particularly the large brewers, were controlling the price of beer at an artificially high level because they both brewed the drink and owned or controlled many of the retail outlets - the pubs. Despite the brewing industry being "probably the best organised political lobby in the country - apart from farmers" (Howkins, 1989), and having always been a steady contributor to Conservative party funds, the Conservative government supported the findings of the MMC Report by reducing the economic power of the large brewing companies. Thus brewers were forced to sell off, franchise (as in the case of Greenall-Whitney) or close half of their tied pubs in excess of 2,000 houses, including giving their remaining pub tenants the freedom to stock 'guest beers' of their choice (Commons Select Committee for Agriculture, 1993). In practice the number of pubs declined, many tenants became lessees or were ousted altogether from their pubs, the price of beer rose, guest beers did not become ubiquitous (*The Economist*, 1991), concentration in the brewing industry continued, and the variation between pubs declined as many publicans tried to reconstruct the 'popular' image of more successful pubs elsewhere (Tomkinson, 1993; Whitebloom, 1991). These changes have characterised both rural and urban areas, but have often been more deleterious to village life where fewer pubs exist.

The sale of pubs resulting from the MMC Report has led to the rise of new independent chains of pubs such as Northumbria Inns (based in Newcastle), J.D.

Wetherspoon, Regency Inns, Enterprise Inns and the Pubmaster chain, which have nevertheless reinvigorated the trade (Gates, 1993; Rawstorne, 1991). These chains have helped the consumer by maintaining a choice between pubs - and thus recreational/entertainment options - and have also aided many pub tenants who were given a new lease. The breweries also gained in that they were able to sell some pubs that were not useful to them, rather than closing them down with the consequent financial loss. Thus the Brewers' Society predicted, too pessimistically as it transpired, that the implementation of the Report would lead to the end of the village pub (Howkins, 1989) and would be "as devastating to the British landscape as the dissolution of the monasteries" (Whitley, 1992). Nevertheless, many pubs have been boarded up, turned into residential houses or demolished, while the new pub-owning companies have sought to develop their trade by establishing popular, homogenised 'house-style' images for the pubs in their ownership.

Consumption of the Village Pub and the Influence of Social Recomposition in the Countryside

The published literature contributes little in helping to define the consumption of the pub in village life; indeed where the pub has been researched, an urban rather than rural context has been investigated (Gofton, 1986; Rogers, 1989; Katovich and Reese, 1987). The 'rural community studies' of the 1940s-1950s in Britain (Bell and Newby, 1971; Lewis, 1986), for example, have little to say as regards the role of the pub. Certainly the pub provided an arena for male-dominated social interaction and the formation of friendship groups, but the segregation of the pub into 'public' and 'lounge' bars suggests that gender and social class divisions in village life were carried into rather than blurred by the pub. More recent studies of rural social change have also tended to ignore the role of the pub (Harper, 1989), with some anecdotal evidence that where a village has more than one pub, 'newcomers' gravitate to one and 'established' villagers to another, thereby underpinning a more recently established social class division within the village (Marsden *et al.*, 1993, pp. 25-26).

Here the role of the pub is contextualised by the changing function of the English rural village. Summarising, a new 'service class' has occupied the village, together with a 'retirement class', thereby displacing the traditional class structure based on landed capital, farm employment and employment in other traditional rural industries and services. The planning process in rural areas (Cloke and Little, 1990; Murdoch and Marsden, 1994) has enabled these changes to act in a socially selective manner through the housing market; social recomposition is most evident in those villages subjected to planning 'restraint' on their housing development, with only the more wealthy classes having access to the restricted number of residences. While seeking entry into the 'rural idyll', the service class has brought new cultural values and behaviours into the countryside, with consequences for the village pub.

To investigate these behaviours and their consequences, a study has been made of one (anonymous) rural village in a lowland county of England, namely Leicestershire. The case study village is judged to be representative of small villages in rural, lowland England: it has a population of 450 distributed over 252 households, supports two pubs (three until recently) and a shop/post office; until the last ten years, the village was subjected to 'restraint' as regards planning consent for new housing and consequently experienced little population growth. The village is within commuting distance of several urban areas, including the city of Leicester, but it does not lie on a major trunk road and has an infrequent bus service and no railway service. A questionnaire survey of all households in the village yielded 55 usable responses (a 30% sample) on the place of the two pubs in village life.

Table 12.1. Frequency of visiting to village pubs (% in each household type)

Household type	At least once a week	At least once a month	More than once a month	Never	% Total Households
Male(s) only	86	0	14	0	13
Female(s) only	63	10	27	0	20
Males and females	39	39	22	0	56
Neither males nor females	0	0	0	100	11
Total	45	24	20	6	-

Source: Field survey by authors

As anticipated, the village is socially polarised between 'newcomers' of the last ten years (51% of households) and 'established' villagers of more than 20 year's residence (35% of households). The former group is dominated by the service class (professional and managerial occupations - 49% of households), and the latter by a miscellany of occupations ranging from farm and forestry workers to a writer and painter. The 15% of elderly single-person households are concentrated in the 'established' villager group, while the 14% of younger single-person households, predominantly male, are found in the 'newcomer' category.

Four groups of households emerged on the use of the two village pubs (see below), cutting across most of these socio-economic characteristics (Table 12.1). Group 1 (13%) consists of households where only the male uses the village pubs. 86% of these males visit one of the two pubs at least once a week and for them the pub continues its traditional role as a social meeting point. This behaviour is not associated with any particular age, socio-economic or marital status group; nor is one pub preferred over the other. Group 2 (20%) is made up of households from which only the woman (usually a single person household) visits the village pubs; 63% of these women visit a pub at least once a week, mainly in the company of female friends or family, but again without any common characteristics of age or socio-economic group. Group 3 - the largest with 56% of households - comprises married couples who both visit the local pubs, usually together. But the majority (61%) visit less than once a week, show a preference for the longer-established pub in the village, and are members of the 'newcomer' class. The lower visiting frequency by this new social group in the village (Table 12.2) results in falling custom for the publican and an increased dependence on visitors from outside the village to maintain the pub's economic viability. In Group 4 (11%) are those households from which no-one visits either of the village pubs. This behaviour is associated with elderly households but no other socio-economic characteristic. For all groups together, the motivation for visiting the pub was equally divided between enjoying an alcoholic drink outside the home, consuming a bar meal (not a full restaurant meal) and meeting friends. On a seven point scale (Table 12.3), 42% of those interviewed considered that the pub played a 'very important' (point 7) role in the social life of the village, but 30% rated it only 'slightly important' (point 5 or less). Villagers in the 'established' class rated the importance of the pub more highly than those in the 'newcomer' class, as reflected in their more frequent use of the pub on a weekly basis.

Table 12.2. Pub visiting by 'established' and 'newcomer' households (% in each type - highest frequency in household)

Household type	At least once a week	At least once a month	More than once a month	Never	% Total Households
Established	53	5	16	26	35
Newcomer	36	39	14	11	51
Total	43	26	15	16	-

Notes: 'Established' defined as resident for >20 years
         'Newcomer' defined as resident for <10 years
Source: Field survey by authors

The picture that emerges is of a changing pattern of use of the services provided by the two village pubs as between the 'newcomer' and 'established' groups. Members of the former group tend to be less supportive of the village pub and its social milieu, preferring to develop their social networks in other ways, often based on previously established social contacts outside the village. Despite moving to a 'rural idyll', it appears that a majority of 'newcomers' prefer to live in rather than be part of village life as measured by their consumption of the village pub. Consequently, despite the high rating given to the social role of the pub in village life by respondents in all four pub-user groups, most 'newcomer' households only participate infrequently and not at a level sufficient to maintain the pub business. As a result, the publican must turn increasingly to the non-village population for the economic viability of the pub business, and adapt the village pub to the consumption preferences of that group.

Table 12.3. The perceived importance of the pub to the social life of the village (% in each household type)

Household type	5	6	7	% Total Households
Established	53	5	16	35
Newcomer	36	39	14	51
Total	43	26	15	-

Notes: 'Established' defined as resident for >20 years
'Newcomer' defined as resident for <10 years
Scale: 1: Unimportant; 7 Very important
Source: Field survey by authors

Consumption of the Village Pub and the Influence of Changes in the Leisure Behaviour of the Urban Population

To explore the influence of the changing leisure behaviour of the urban population, a questionnaire survey of customers was carried out in the two pubs of the study village over two summer week-ends, covering the full range of opening hours. Customers were sampled at random and 87 usable questionnaires were completed (approximately 20% of all customers). One of the pubs is a free house owned by the publican, with parts of the structure over 400 years old. With its thatched roof, this pub has the 'traditional' external image of an English village pub; but internally it contains a restaurant section set within a modern 'open-plan' room structure. The other pub, of more recent origin, is over two hundred years old and is tenanted from one of the large Midland breweries. It is brick built with a tile roof but has a large space adjacent to the car park for tables and a play area for children. Internally there is a restaurant space, but with the area subdivided into smaller rooms in the traditional style.

There are a number of differences in the consumption of the two pubs. The older pub, with its thatched roof and outward 'traditional' appearance, tends to attract the visitor travelling a longer distance, especially the first-time visitor, and the more affluent, professional, older customer. The younger pub, with its cheaper food prices and space for children at play, serves a less wealthy, manual-worker, younger clientele and those who make more frequent visits. The publicans, aware of these differences in consumption, cater for the popular tastes of their clienteles by varying their menus, internal decor and openness to the presence of children. The latter development is relatively new for the English pub. As the village pub is adapted to the popular culture needs of the majority of its customers, so its attractiveness as a space for social interaction by the village population is decreased and its changed role reinforced.

Taking the survey data for the two pubs together, the limited role of the village pub for social interaction amongst village people is confirmed. Only 23% of the customers were drawn from the village, and only 13% were visiting the pub to meet friends who live in the village. On the other hand, there were no differences between the pubs in these features, so that a 'newcomer/established' villager division was absent. Rather, the survey confirmed the more important role of the pub in the leisure behaviour of the non-village population: 30% of customers were visiting with friends from outside the village and two thirds ate outside the home at least once a week; 24% were in the pub specifically to consume food rather than drink. Also the non-village customers were drawn from a wide spatial range: 39% from the city of Leicester (45 minutes away by car), 42% from other nearby towns and 19% from surrounding villages. For 24% of customers, it was their first visit to the pub ('first tasters' in the marketing jargon of the brewing companies (Davidson, 1998)), but a majority of customers reported that they had increased the frequency with which they visited pubs in the last five years, as well as a greater variety of pubs. The picture to emerge, therefore, is of the village pub serving the recreational/entertainment needs of a mobile, discerning, non-village population, with the provision of food in addition to drink having a greatly increased importance. Customers now cover all age ranges (38% were under 30 years old and 34% were over 50) and socio-economic groups (25% had professional occupations and 24% were manual workers).

Thus one of the main changes in the consumption of the village pub is the increasing use made of it as a recreational/entertainment service by the non-village, usually urban population. Indeed, the opening up of the pub to non-village influences lies parallel to the opening up of the village social system to the influence of the new service class. Thus consumption by the new service class, together with the increased dependence of the pub on customers from nearby urban areas, have supported the restructuring of the village pub. Equally significant, however, is the way in which the consumption of the pub reflects changes in the role of women in society and their increased involvement in the leisure activities of the family, especially jointly with their partner. 45% of the randomly selected customers in the survey were women: 13% were visiting with their children and 28% with their spouses/partners. Groups of young women, as well as groups of young men, were represented in the survey showing how the village pub increasingly serves the same functions as its urban counterpart.

Conclusion

This paper suggests that a broader understanding of the restructuring of rural services can be obtained by exploring the processes of production and consumption. In this study of the English village pub, four interconnected processes have been considered: (a) the investment policies of national and international brewing companies; (b) recent UK regulations for the brewing industry; (c) changing social behaviour of the new rural service class; and (d) change in the leisure behaviour of the non-rural population. Assuming the case study village, and its two pubs, in the survey is representative of most lowland English villages - and there is no reason to judge that it is not - the pub can be shown to have a significantly altered place in village life. The pub is playing a less important role in the social network of a majority of villagers, especially in 'newcomer' households (Table 12.2); nevertheless, a perception of the importance of the pub remains for 'established' households (Table 12.3), while single men and women still retain the village pub as an arena for social interaction. However, the role of the pub is now much more concerned with the production and consumption of a recreational/entertainment service for non-rural people, as underpinned by the capital accumulation demands of the brewing companies and the new pub-owning companies. These companies have a view of the popular culture needs of the urban population and are adapting the physical structure and services of the village pub to serve that perceived urban need.

Chapter 13

Competitive Tensions in Community Tourism Development

Alison M. Gill

Introduction

Tourism developments have become integral components of many rural systems that have undergone economic restructuring. It is now widely recognised that the impacts of tourism extend beyond simply economic considerations to social and environmental concerns, but the complex dynamics of the tourism system and the manner in which it is integrated into community and regional systems remain weakly theorised (Britton, 1991; Ioannides, 1995; Ioannides and Debbage, 1997). Tourism is generally categorised as a component of the service sector economy, but at the same time it is also a resource-dependent industry (Innskeep, 1991). The need for community involvement in decision-making regarding tourism development is seen by many researchers (e.g. Murphy, 1985; Blank, 1989; Keane, 1990) as essential to the success of the industry. However, there appear to be limitations, or at least major challenges, to planning for tourism at the scale of the community.

In this paper, a case study of a forest resource-dependent community in British Columbia is presented to examine the process of community-based tourism decision making and to explore the competitive tensions that arise as the result of this process. Recent development approaches have shifted much fiscal and decision-making responsibility to the community level, thus, communities are experiencing not only economic change, but also changes in local political structures. The intent of this paper is to gain a clearer understanding of the challenges posed to emergent community development approaches by the inherent nature of the local tourism production system. The paper first presents some key elements, drawn from the literature, on the relationship between the community, the tourism production system and local economic development processes. The subsequent case study of Squamish, British Columbia reveals examples of competitive tensions that emerged during the process of planning for community tourism. The paper concludes by discussing these observations with reference to findings elsewhere.

The Community, the Tourism Production System and Local Economic Development

The following definition of 'community' proposed by the sociologist, Roland Warren (1977, p. 208) seems especially appropriate to the discussion of tourism and the community as it conceptualises change as an outcome of competition. He defines 'community' as:

> "an aggregation of people competing for space. The shape of the community, as well as its activities are characterised by differential use of space and by various processes according to which one type of people and/or type of social function succeeds another in the ebb and flow of structural change in a competitive situation." (Warren, 1977, p. 208)

The transition within a community from one form of economic endeavour to another in which there are shifts of political and economic power has been demonstrated to be a contested process (Marsden and Flynn, 1993; Reed, 1997; Reed and Gill, 1997). In this paper the focus is on the tensions that emerge with the introduction of tourism into a rural community and that lead to competition in various arenas: political; economic; social; and environmental. A fundamental difference between tourist places and other communities is that residents compete with tourists for basic community resources such as space and facilities. There is an essential dichotomy between residents who view the community as their home and a place to live and the tourist who views the community as a resource that is commodified and consumed (Britton, 1991; Urry, 1995). Small communities are especially affected as few parts of the community are exclusive to residents. The extent and nature of the tourism production system, which is defined as "the various commercial and public institutions designed to commodify and provide travel and touristic experiences" (Britton, 1991, p. 455), is place specific. Regardless of the specific type of tourism there are several key characteristics that present challenges to community decision-makers and distinguish tourism from other economic activities (see Table 13.1).

One of the main challenges is the fragmentation of the tourism production system. While from the perspective of the tourist, the community is experienced as a single product, it is produced by a diversity of producers. In turn these producers are dependent on tourism to varying degrees. Such fragmentation not only makes it difficult to estimate the true contribution of tourism to the community economy, but also makes it hard to ensure quality control. Further, fragmentation of the industry also means that there is no common voice for the industry as a whole, indeed there is frequent competition between producers within the industry

(Debbage, 1990). This weakens the power of the industry to compete for resources against more unified sectors of the economy.

From an environmental perspective, conflicts may arise because of the multi-jurisdictional nature of the resource base upon which tourism is dependent. Especially in rural areas where outdoor recreational activities and scenic resources form the basis of the tourism industry, the resources often lie beyond municipal control. Keane (1990) notes that problems associated with rights of access and ownership can cause a particular form of market failure, which can only be resolved by some form of collective initiative. This may involve some form of private/public sector co-operation. Tourism is also generally dependent on the state for the provision of highways, airports and other infrastructural elements as well as for co-ordinated marketing and place promotion (Britton, 1991).

Table 13.1. Selected characteristics of tourism that challenge community planners

---
A. The tourism product is mainly an intangible experience which is produced as it is consumed.
B. The tourism experience is viewed as a single product but is produced by a diversity of producers - some outside the community.
C. Almost all community goods and services are consumed to some degree by the tourist.
D. The tourism industry is hard to classify as a diversity of producers offer differing proportions of goods and services to tourists.
E. The industry is fragmented. It contains a high proportion of small entrepreneurs and there is often no common voice.
F. There is a high level of private/public sector dependency.
G. Tourist demand is uneven with seasonal peaks.
H. Management of the resource base is often multi-jurisdictional and operators frequently do not control the resources on which they are dependent.
I. The market is spatially dispersed and demand is hard to control.
J. Tourism is perceived to exhibit low productivity levels and depend heavily on low pay, low skill, occasional and female-oriented labour.

Source: Based on Blank, 1989; Britton, 1991; Ioannides, 1995

While there are many problems associated with co-ordinating and controlling the tourism supply system at the community level, the problems associated with predicting and controlling demand are even greater. This is primarily due to the spatial dispersion of the market. The problems of seasonality exacerbate the situation and in many instances make dependency on tourism, as the major economic enterprise within a community, problematic (Baron, 1975). Seasonality also contributes to the frequent criticism that tourism creates only low paying, low skilled jobs (Ioannides, 1995).

Planning for tourism is a relatively recent phenomenon, primarily undertaken by the state at the national or regional level and with a focus on marketing and promotion. Tourism development has often been the result of private initiatives and has frequently lacked co-ordination. Individual tourism companies have been reluctant to engage in co-ordinated marketing and place promotion because rival firms, who have not shared in the marketing costs, may also reap the benefits (Britton, 1991). Over the past decade planning for tourism at the community level has been more widely advocated (Murphy, 1985; Blank, 1989; Keane, 1990), although not so widely practised.

Interest in local tourism planning is coincident with the general shift to bottom-up development approaches and an element in many efforts to diversify local economies. Local development approaches can, however, take a variety of forms. An important distinction to make is between any form of local economic development (LED) which is locally initiated, but may still be controlled by a local elite as opposed to a community economic development approach (CED) which is process-oriented and has social as well as economic objectives. Increasingly, as the result of the gradual retrenchment by senior governments from regional development, local development efforts are confined within municipal boundaries (Filion, 1991; Bryant, 1995). Cox and Mair (1988) argue that current LED approaches increase inter-jurisdictional competition and that confrontation at all levels of government is exacerbated. Inter-municipal co-operation is precluded because of the absence of joint gains as local government institutions are locally dependent and are interested in local economic growth. Elements of these problems will be demonstrated in the case study. Collaborative approaches are thus made more difficult as local governments are forced to compete for increasingly scarce fiscal resources, for example in a tourism-related context, for such projects as downtown revitalisation or heritage restoration.

The Case Study

Squamish is a District Municipality in the province of British Columbia. It is located 66 km north of Vancouver at the head of a deepwater inlet, Howe Sound (Fig. 13.1). Like many resource-dependent communities in British Columbia, its economy has been based primarily on forest industries - logging, pulp production, saw milling and other log handling activities all of which have been affected by restructuring (Barnes and Hayter, 1993). While, unlike some coastal operations, the pulp mill and sawmilling operations in Squamish have not closed down, they have experienced downsizing. Despite these employment losses, the population of Squamish has grown. This is primarily due to demand for affordable housing within commuting distance of Greater Vancouver. The amenity value of Squamish is increasingly being recognised as it is located along a scenic coast route to the destination resort of Whistler, in what is fast becoming a diversified four-season recreational amenity corridor with several golf courses, high quality rock climbing, windsurfing and other adventure tourism resources. The population is now estimated to be around 13,000 and growing at a rate of at least 3% per year. The change is not simply numerical: there is also a shift from blue collar to white collar workers and, thus, a significant proportion of the workforce is not dependent on the community for employment.

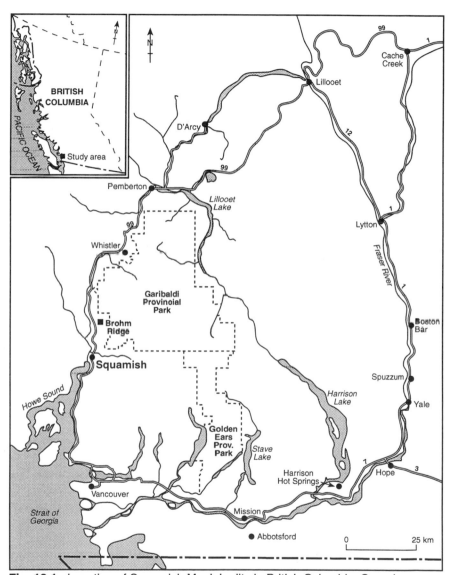

Fig. 13.1. Location of Squamish Municipality in British Columbia, Canada

The Community Tourism Development Process

Until recently the only tourism developments that were seriously considered by the local municipal council were large-scale exogenous developments that would enhance the community's tax base. Such an approach to local economic development has been termed by Bryant (1991) the traditional "industrial development model". One such project that gained the recent support of the council was the development of a major ski hill, Brohm Ridge. This proposed development is located outside the municipal boundaries but it was supported because of perceived employment opportunities and the likelihood of spin-off benefits to the community, such as a boost for Squamish's image. The provincial government's refusal to support the project, in part due to the fact that the proponents had not sought public approval from the community, was the stimulus for the community to engage in a community tourism planning process.

In 1992 Squamish announced it would undertake a Community Tourism Action Plan (CTAP). This is a provincial government programme in which the Ministry of Tourism acts in a facilitating role to help communities make decisions concerning their tourism development. The first step was the establishment of a Steering Committee which was composed of appointed representatives from municipal and regional government agencies as well as from the Chamber of Commerce, First Nations and BC Rail. The role of this initial committee was to gather whatever community data were available on market characteristics and identify a preliminary list of assets and concerns relating to tourism. This information then formed the basis for a public workshop held in January 1993, which was facilitated by the Ministry of Tourism, at which participants expanded upon the information and identified objectives and priorities for action.

At this workshop, volunteers were sought to serve on a Citizen's Advisory Committee, which had a mandate to make recommendations to the District Municipality in the form of a strategic plan. The twenty residents who volunteered were assigned to work on one of two committees, depending on their interests. One committee dealt with promotional issues, marketing and community education, while the other was oriented towards product development of attractions, infrastructure and services. Both of these groups received facilitation assistance and resource support from the two economic development agencies operating in the municipality, the Sea to Sky Economic Development Commission and the Howe Sound Community Futures Society.

After months of discussion little progress had been made by either committee towards the goal of developing a tourism plan, so in July 1993 the committees regrouped into a single Citizen's Advisory Committee in order to establish a vision statement. The statement that they eventually reached consensus on read:

"to build and strengthen a diverse four-season tourism sector while maintaining our small town atmosphere and preserving our heritage."
(District of Squamish, 1994, p.5)

With this in place and with the Community Futures Society acting as the lead agency, the committee set to work to develop a plan. In September 1993 a Tourism Co-ordinating Committee was formed as an expanded version of the original Steering Committee including more representatives of government agencies and major land owners in the Squamish Valley. The lack of discussion of the Brohm Ridge ski development, which had been the original reason for initiating the process, led to the creation of a separate Winter Tourism Development Committee which functioned parallel to the Citizen's Advisory Committee. The plan went through several draft and revision phases and was finally adopted in principle by the Municipal Council in December 1994. The Steering Committee and the Advisory Committee were both disbanded after presentation of the plan, thus there is no individual or agency responsible for advancing the proposals. Subsequently, there has been no further action from the Council.

Table 13.2. Top 14 ranked tourism development objectives, Squamish *Tourism Development Plan*

Rank	Objective
1.	Identify and market an inventory of small business opportunities and services required to support the *Tourism Development Plan*.
2.	Finalise a comprehensive, community-based *Tourism Development Plan* for the District of Squamish.
3.	Develop a plan to promote outdoor winter tourism opportunities and attractions.
4.	Examine ways to support the arts community in the District of Squamish.
5.	Support Squamish Nation heritage, history, culture and research.
6.	Organise and co-ordinate hospitality training and local 'Spirit' programme.
7.	Identify an inventory of assets (and potential tourism attractions) associated with special skill, unique activities and ceremonies of distinctive ethnic groups.
8.	Ensure the necessary land base is available to support the vision and objectives outlined in the *Tourism Development Plan* through appropriate Zoning Bylaw and Official Community Plan designation.
9.	Encourage increased tourism co-ordination.
10.	Develop a public relations plan for tourism in Squamish.
11.	Review permanent cruise ship terminal-activity for the Squamish port.
12.	Encourage partnerships with Squamish Nation on tourism projects and promotional strategies.
13.	Encourage 'grass-roots' public participation in tourism planning and implementation.
14.	Encourage the co-ordination of forestry tours.

Source: District of Squamish, *Tourism Development Plan*, 1994, p.8

In the *Tourism Development Plan*, thirty of the 103 objectives identified at the initial public workshop were further developed by the Advisory Committee into generalised action plans. In addition, the Winter Tourism Committee separately proposed a further six objectives which were incorporated to make a total of 36. Of these, 14 were recommendations for specific projects, but the remaining 22 were associated with research and planning, logistical support, training, co-ordination and infrastructural development. When these objectives were ranked by the Advisory Committee according to priority (see Table 13.2), the top-ranked specific development project - the development of a cruise ship terminal - was ranked eleventh. The Brohm Ridge ski development was not specifically identified but was subsumed under the third-ranked objective of developing winter tourism options.

Discussion

The process of tourism planning in Squamish - the community's first experience at engaging in a community development process - reveals a number of competitive tensions that arose as the community tourism planning process challenged the interests of traditional elites (see Table 13.3). These tensions were revealed in political, socio-psychological and economic arenas at both the intra- and inter-community levels.

Table 13.3. Competitive tensions in the Squamish Community tourism planning process

Arena	Intra-community tensions Issue	Example
Political	CONTROL	Traditional elite vs. community involvement
Social/Psychological	REPRESENTATION/ VALUE	Long-term vs. short-term residents
Economic/Environmental	RESOURCE ALLOCATION	Sectoral competition (e.g. forestry vs. tourism)
	Inter-community tensions	
Political	FISCAL RESOURCES	Municipal competition (e.g. heritage development funds)
Social/Psychological	IMAGE	'Us' vs. 'them'
Economic/Environmental	MARKET COMPETITION	Tourism market

Intra-community Tensions

An obvious and central tension that emerged in Squamish, is one common to communities undergoing transition from top-down to bottom-up decision-making processes, namely, the tension caused by the differing value systems of the traditional elite compared with the general public (e.g. Ethos Research

Associates Inc., 1995). In the specific context of tourism, this tension was most obvious with respect to the reactions of the Chamber of Commerce. In Squamish, as elsewhere, the Chamber represents the traditional business elite of the community and their role is frequently that of boosterism. A representative of the Chamber was actively involved as a member of the original Steering Committee, but withdrew from the process when it became clear that the construction of a large hotel was not a high priority of the Citizen's Advisory Committee. The importance of establishing a large hotel as a community image booster is a frequently favoured tactic, even in large cities (Britton, 1991). Later in the process, the Chamber was asked to review the draft planning document and the resulting comments led to the alteration of two objectives. Both of the recommendations to which the Chamber objected were related to the hiring of individuals to be responsible for tourism co-ordination and monitoring. Resistance by Chambers of Commerce to community-based tourism planning efforts is not uncommon. In the case of Squamish it appears to be less a matter of fear of competition to existing tourism businesses, as documented in other cases (Blank, 1989), and more a fear of loss of control over tourism initiatives in the community. However, subsequent to the acceptance of the tourism plan by the Council, the Chamber of Commerce has been officially reaffirmed as the agency responsible for tourism.

The other obvious tension between the traditional elite approach and the community-based process was with respect to differing attitudes towards the Brohm Ridge ski resort proposal. While the Brohm Ridge proposal was the reason that the local council initiated the process, it quickly became lost as a priority of the Advisory Committee. It was only after urging by the ski developer and council supporters that interest in the project was resurrected with the creation of the Winter Tourism Advisory Committee. This committee acted in a parallel fashion to the main advisory committee and their recommendations were only incorporated into the final planning document after some debate, although the specific proposal was merged with more general winter tourism objectives. As demonstrated by the priority rankings of objectives for tourism development, the community tourism planning process in Squamish reinforces findings elsewhere (e.g. Keane, 1990; Bryant, 1995) which demonstrate that community-based development processes have broader agendas than simply the development of economic projects and encompass social and environmental issues such as concerns over quality of life, sustainability and the development of human capital.

Tensions in the tourism planning process also emerged around issues of representation. The most notable feature of the Advisory Committee's composition was the length of residence of participants. Of the 20 members only two were long term residents with over 20 years experience. Over half (11) had lived in the community for less than five years and a total of 16 for less than seven years. Differences in values between newcomers and established residents have been documented in many contexts (e.g. Blahna, 1991; Cloke and Little, 1990). As Blahna (1991) found in a study of forestry communities in the US Pacific North-West, conflicts do not only result from differences in the culture of newcomers and long-term residents, but from the way newcomers are integrated into the social and political life of a community. The high rate of participation by recent residents in the tourism advisory committee in Squamish suggests that newcomers find voluntary citizen's advisory groups a readily accessible form of participation in community decision-making.

While some of the newcomers participating in the Advisory Committee had entrepreneurial interests relating to tourism, others were simply interested citizens. Newcomers were much more insistent on including the term "small town atmosphere" in the vision statement than longer-term residents. The visioning process and the resultant broad statement was useful in diffusing original project specific ideas and subsequently refocusing attention on a truly community development approach, in which the objectives were not simply economic but representative of broader quality of life issues. The inclusion of terms such as 'small town atmosphere' and 'heritage' echo similar visions of rurality in other settings. As Cloke and Little (1990) observe, these small town values, which embody notions of community spirit and overall quality of life, are common to rural middle class residents in the UK.

The composition of the Advisory Committee also raises questions concerning sectoral representation. The multi-jurisdictional nature of tourism-based resources means that tourism interests need to collaborate with other resource stakeholders to reach consensus over such issues as access and scenic preservation. Both these issues are extremely pertinent to the tourism resource base in Squamish, and co-operation, especially with the forest industry, is an important key to maintenance of landscape aesthetics. The absence of representation from the forestry sector is therefore of particular concern. While a member of the Ministry of Forests was initially invited to sit on the Steering Committee, involvement was not sustained. Furthermore, no members associated with the forest industries sat on the Citizen's Advisory Committee. From the perspective of some Advisory Committee members, this lack of representation was not seem as a problem as they did not perceive the role of committee members as that of 'sectoral representative' but rather 'concerned resident'.

However, some attempts were made by the Advisory Committee to address obvious gaps in representation. For example, when First Nations participation became sporadic, a special meeting was arranged to elicit input into the process. Co-operation with First Nations is a critical current political issue in Canada, and with respect to tourism development in the Squamish region is relevant to the development of a cultural tourism product and to issues of recreational access. The inclusion of the broad term 'heritage' in the vision statement embodied the notion of cultural heritage, especially First Nations culture, while at the same time encompassing all other heritage domains including the natural environment, the traditional forest and railway industries and the resident Sikh population.

Further, in an attempt to gather feedback from groups not represented on the committees, the draft tourism plan was circulated for comment to a broad group of stakeholders. Representation is linked to empowerment in participatory processes (Shragge, 1993) thus, inadequate representation and support from key sectors within the community could impede implementation by allowing traditional elites to argue that the views expressed are not fully representative of the whole community.

Inter-community Tensions

Because of the municipal context of the planning process, the most obvious tensions in the Squamish tourism planning process were intra-community issues. However, inter-community tensions were evident during committee discussions and are likely to become more pronounced in the event of attempts to implement certain development recommendations which position the community competitively with respect to other local communities in obtaining public funds. For example, infrastructural support for such developments as the cruise ship terminal, would require a major commitment of funds from the public sector. A long-standing tension exists between Squamish and Whistler regarding public sector spending, particularly with respect to highway expenditures. Squamish residents perceive bias in favour of projects that benefit Whistler at their expense. The competitive relationship with Whistler was also evident in discussions relating to the vision statement for Squamish. The debate surrounding the 'small town atmosphere' frequently reverted to the desire not to become like 'Whistler' but to remain a 'real' community instead of the artificial resort that many perceive Whistler to have become. Whistler was characterised as an example of the negative impacts of tourism.

The 'us' versus 'them' debate also emerged with respect to market competition with Whistler. Squamish is one stop, for example, on what the provincial Ministry of Tourism is prominently marketing as the Sea-to-Sky Circle Route, a 700 km touring loop beginning and ending in Vancouver. Such exogenous market influences make municipal control difficult. The opportunity to 'capture' some of this potential market at the expense of Whistler and other local communities was a component of the discussion of marketing issues. The Brohm Ridge ski proposal would also be in direct competition with the Whistler ski market, offering an intervening opportunity for those travelling from Vancouver. A major dilemma for any rural community tourism planning lies in the fact that a high proportion of rural tourists is actually touring. Thus, the tourist experience is not that produced by any single community, but rather several communities within a region. This leads to dependency relationships that are regional rather than municipal.

Conclusions

Much community-level tourism development has been *ad hoc*, driven by traditional market forces. Recently, however, in concert with emerging bottom-up processes of community development, municipally based tourism planning is becoming increasingly advocated. In this paper, the competitive tensions that emerge during such a process have been examined using a case study of Squamish, British Columbia. The aim has been to uncover some of the problems inherent to the tourism production system which make community-level planning more challenging. The findings bring some convergence to a variety of issues that have emerged in other discussions on rural community change.

While the findings are derived from a case-study and are therefore only pertinent to the specific locality, some observations may prove relevant to other contexts. The biggest challenge in the Squamish citizen planning process seemed to be the relationship between representation and empowerment. The tourism sector, unlike other resource based industries, is faced with the challenge of not controlling the land base on which it is dependent. Co-operation and support from the sectors who do control the resources is necessary if citizen-based tourism recommendations are to be implemented. However, it seems difficult to get support from such sectors either because of perceived land-use conflict (e.g. clear-cuts versus viewsheds) or disinterest in tourism initiatives. Likewise, established community elites such as the Municipal Council and Chamber of Commerce, have agendas that are embedded in more traditional approaches to local economic development. While the process in Squamish demonstrated that the community-based process was effective in diffusing pressures to focus on exogenous development options and instead resulted in an emphasis on capacity building, empowerment to follow-up on recommendations still rests with the established elite. This regression to the norm of governance led one committee member to describe the process as 'the big fizzle'. Some argue that the success of a citizen involvement process cannot simply be measured in terms of economic outcome but also in terms of the value of the process itself in stimulating debate and heightened awareness (Cloke and Little, 1990; Keane, 1990). In Squamish, this is to some extent true and individual members of the committee have gained experience and knowledge.

The role of newcomers in the community tourism planning process is also notable (Gill and Reed, 1997). Many of the newcomers active on the Advisory Committee saw their role as that of 'interested citizen' and had no links to the tourism industry. This seems to suggest the common ground between the interests of the tourism sector and those of the resident. Both are interested in amenity issues such as recreational access and scenic resources. There are few barriers, other than the commitment of time and energy to voluntary processes such as the Citizen Advisory Committee in Squamish. However, over-representation from this sector may also be a factor of resistance in the acceptance of recommendations.

In conclusion, due to the fragmented nature of the tourism production system and the invasiveness of tourism in small rural communities, there seems little question that the development process should be integrated with a community vision, which embodies essential quality of life values for residents. However, as community tourism planning increasingly moves beyond strategic planning towards product development, collaboration rather than competition is necessary to ensure complementarity of product and embed local efforts in larger regional scale plans.

Chapter 14

Democratising Rural Development: Lessons from Recent European Union and UK Programmes

Steve Martin

Introduction

This chapter explores the increasing emphasis on community involvement in rural development initiatives funded by the UK government and the European Commission. It examines the ways in which community involvement has come to be seen as a key ingredient in rural policies and how this has been reflected in the activities of a range of agencies at European, national and local levels. It then presents the key findings of evaluations of two of the most important British and European rural development initiatives of recent years. The conclusion offers an assessment of the insights that these programmes provide, into the potential of, and challenges facing, attempts to promote increased community involvement in rural policy-making.

Community Involvement Moves into the Mainstream

Over the last two decades a variety of initiatives in the UK and the EU have aimed to increase the involvement of local communities in the design and implementation of rural development programmes. The philosophies underlying these approaches draw upon on a number of traditions on both sides of the Atlantic. Increased interest in public participation among policy-makers can be traced back to the late 1960s and early 1970s. In particular the Skeffington Report (Skeffington, 1969) and Alinski (1972) were influential advocates of community consultation as a means of combating the alienation from the policy-making process felt by many 'ordinary people'. In the UK, though, it was in the 1980s that the value of promoting local action really began to be recognised across the political spectrum.

Central government, strongly influenced by the Right, advocated reliance on market forces as opposed to the state, but also encouraged a strengthening of civil society as an important means of combating the dependency culture, which it believed had been spawned by excessive government intervention in the post War period. So called 'active citizenship' was advocated as a means of increasing individual choice and shifting control from public sector professionals to policy consumers. The requirement that local communities should be actively engaged in proposed regeneration projects as a prerequisite of central government funding was one of the more tangible expressions of this. More recently community involvement has come to be seen by central government as one of the main vehicles for achieving commitments it made at Rio, and subsequent 'Environment Summits'. Community-led action has also attracted increasing support from some sections of the Left - in particular from a number of local authorities which have decentralised service delivery to neighbourhood offices and sought to consult with local people through the establishment of a variety of new neighbourhood fora (Burns *et al.*, 1994).

Alongside these developments at central and local levels, both the private and voluntary sectors have also sought to gain legitimacy by working more closely with local communities. A growing number of voluntary agencies have claimed a role as the 'voice' of local communities and a wide range of businesses has embraced notions of corporate citizenship. In some cases the private and voluntary sectors have come together to provide technical aid to communities through for example 'Groundwork Trusts' and the national Shell Better Britain Campaign.

Community involvement in regeneration and development initiatives has also become an increasingly important element in EU policy and a requirement of two of the most important of its Community Initiatives - URBAN which is targeted on deprived inner city areas and the LEADER programme which seeks to promote a 'bottom-up' approach to rural development initiatives across Europe.

These trends have been influenced by and reflected in a range of 'self-help' initiatives in rural areas (Wright, 1992). Rural local authorities have, on the whole, been less actively involved in experiments to engage with the local communities. This, and the relative paucity of state funded social and physical infrastructures in some parts of rural Britain, has meant that community action has often become a necessity and in many areas local people have organised schemes such as community transport provision to ameliorate the worst effects of the withdrawal of public transport. Moreover, unlike urban areas, most of non-metropolitan England is served by a network of elected parish councils, the best of which have acted as important foci and catalysts for local action with many villages benefiting from a strong sense of local identity, which has made it possible to generate a high level of community involvement.

Government has sought to stimulate and also to tap these efforts. In particular its three main countryside agencies, the Rural Development Commission (RDC), the Countryside Commission and English Nature, have increasingly stressed the importance of local action. The first two have a long history of support for community led initiatives. The RDC has played a key national role in highlighting problems of rural deprivation, the lack of affordable housing, threats to local services and social exclusion. It has sought to address these through funding for

Rural Development Programmes (Martin *et al.*, 1990) and a network of voluntary organisations - Rural Community Councils. Through these agencies and programmes it played a key role in encouraging 'local appraisals' in hundreds of parishes throughout rural England during the last twenty years (Bowler and Lewis, 1991; Sherwood and Lewis, 1994), and provided support to village services and sought to improve access to advice and information, leisure activities and health care (Bovaird *et al.*, 1992). In recent years it has also funded community development workers in a number of the most deprived areas (Tricker and Martin, 1990).

The Countryside Commission has placed a similar emphasis on local action since the mid-1980s when it funded a 'Community Action Experimental Programme' to test out a range of local environmental pilot schemes. Its 1990 policy statement emphasised the importance of local action (Countryside Commission, 1990) and was followed by new community-led schemes, including the Countryside Initiative and the Community Woodlands Scheme. The Commission was also instrumental in setting up a steering group, comprising officers from the Countryside Commission, the Shell Better Britain Campaign (SBBC), the Royal Society for Nature Conservation (RSNC) and the British Trust for Conservation Volunteers (BTCV), which began to develop a national countryside environmental action scheme upon which the Rural Action for the Environment programme (see below) was modelled.

English Nature (and its predecessor the Nature Conservancy Council) has a shorter history of involvement in community-led schemes. However, from the early 1990s onwards it has stressed the importance of a 'people-orientated' approach to the safeguarding of natural habitats (English Nature, 1993) and introduced a range of new locally led initiatives including the Community Nature Scheme, Community Action for Wildlife and the School Grants Scheme (English Nature, 1991; Millward, 1995).

The most recent White Paper on Rural England (HMSO, 1995) endorsed these developments within all three agencies. It reiterated the Government's commitment to encouraging local initiative and voluntary action as a means of identifying needs and solving local problems stating that people's willingness to help each other is "one of the great strengths of rural England". The Government's aim was therefore "to work in partnership with local people rather than impose top-down solutions" and in this context "Mechanisms such as village appraisals and local housing needs surveys can help communities to define their priorities, identify what they can do to meet them and target limited resources effectively." In support of this view the White Paper cited the success of the first three years of its Rural Action for the Environment initiative in encouraging and supporting voluntary efforts to improve local environments.

Rural Action for the Environment

The Rural Action for the Environment initiative was a major component of Government's 'Action for the Countryside' programme. Launched following the 1992 General Election, the programme had a modest initial budget (£3.2

million over three years) provided equally by the Countryside Commission, English Nature and the RDC. It was overseen by a national Steering Group comprising senior officers from these agencies (the so called 'sponsors') and five national voluntary organisations[1] each of which had strong links with one or more of the sponsors. A small (four person) National Development Team was created to implement the initiative.

Rural Action (as it became known) incorporated many of the strands of what was regarded as 'good practice' in rural development. In particular, it was seen as a test bed for an enabling/empowering model of rural development in which local people took the leading role in developing and implementing policies for their areas rather than acting as the passive recipients (or victims) of 'top-down' policies devised and delivered by professional experts. This reflected commitments to local action made by central government in the Environment White Paper (Department of the Environment, 1990) and the Rio Summit Declaration (United Nations, 1992; Local Government Management Board, 1994). Thus, as the minister of the day put it, "Rural Action is not government and its agencies standing aloof from those whose lives it will affect. It is local people themselves who will act to improve their environment, tackling the problems they have identified, in the way they consider best" (Department of the Environment, 1992).

Support to local communities was provided by partnerships of key rural development and environmental agencies in each county. These became in effect self-governing networks with a high level of autonomy to organise their activities in ways which reflected local priorities. Most had between 30 and 40 members. However, some of the smallest (for example Cleveland, Nottinghamshire, Isle of Wight, Avon and Shropshire) had fewer than 15 members and the largest (for example Cambridgeshire, Hertfordshire and Bedfordshire) had more than 100 members. Networks were, in theory, partnerships of equals. In practice though most were initiated by a small group of officers from local councils, the BTCV, wildlife trusts and RCCs. These then drew in other agencies.

Each network formulated a 'Network Plan' outlining the ways in which it would promote Rural Action within its county. Once this was approved by the National Development Team, the network became eligible for an annual 'network grant' (on average about £8,000 per annum) to enable it to implement its plans. Organisations wanting to join a network were required to have relevant expertise to offer to local communities and to sign a formal commitment to the principles of the initiative. Nine networks were established in 1992, a further nineteen 1993 and seven in 1994. By the end of 1995 there were 40 covering the whole of non-metropolitan England.

In addition to publicising Rural Action, networks were responsible for advising local groups and visiting potential projects (an activity for which they received no additional funding). Grants of up to £2,000 to local groups were available to fund a maximum of 50% of the total costs of a project. Applications were appraised by the network members and grants were administered by the RCCs, which received a fee based on the number of payments processed. Local groups were required to provide 'matched funding' but this could include the notional 'cost' of volunteer time as well as other contributions 'in kind' - a major innovation designed to emphasise the value of community participation and input. The average size of grants made in the first three years was about £850. One-third of applications were from local amenity/conservation groups, 25% from parish councils, 20% from community-based residents groups and 15% from churches, youth groups, historical societies and other local groups.

By October 1996, four years after the first county support networks began operating, project grants totalling £2 million had been given to more than 2,500 projects and well over 1,000 organisations were reported to be working together through the networks. By this stage it was estimated that nearly 100,000 local people had taken part in projects supported by Rural Action, two-thirds of whom had not previously been involved in working to improve their local environment (Rural Action National Development Team, 1996).

A comprehensive evaluation of the first three years of the programme (Bovaird *et al.,* 1995) examined in detail 200 of the first 600 local projects to receive assistance and the operation of the first 22 county networks to be established. The study concluded that the initiative had been successful in funding local projects which would not otherwise have gone ahead and that most (82%) were considered to have been successful by local people. In many cases schemes had involved active participation by local volunteers with no previous experience of environmental action and nearly all had raised local awareness, increased local peoples' confidence and enhanced their skills in a variety of ways (Martin, 1995). Community groups praised a number of the key elements of the initiative including in particular the county level administration of grants, which enabled a rapid response to requests for assistance and helped them to sustain local enthusiasm for projects. They also regarded the ability to include the value of volunteer time as 'matched funding' as a critical success factor and many welcomed the fact that, unlike most other initiatives, Rural Action could provide repeat funding to enable communities to maintain or extend projects.

The formation of county networks had improved and strengthened the infrastructure of advice, training and support available to local communities and promoted greater joint working between agencies. Only 10% of network members believed that their county's network was not working well and almost two-thirds reported that the number and usefulness of both formal and informal contacts between agencies had increased. More than 60% believed that their knowledge of the activities and priorities of other organisations had improved and 25% that working relationships between network members had been strengthened. In particular, Rural Action was seen as having broken down barriers between agencies concerned primarily with social/community issues (especially RCCs) and environmental agencies (e.g. BTCV and Wildlife Trusts).

There were though some difficulties. Firstly, the regulations regarding the types of projects which were eligible for grant aid were incomprehensible to many agencies and local communities. The complexity of the criteria arose from the vagueness of the initiative's stated objectives and the very broad definition of environmental action which encompassed any activity which "improves, enhances, protects or promotes enjoyment of all aspects of the natural and human environment" (Rural Action National Steering Group, 1993). This was the result of a need to embrace the different agendas of each of the sponsors and partners but meant that the initiative became over-bureaucratised and in some cases, rather than empowering local people, reinforced their dependence on professional experts to interpret the rules.

Secondly, the national sponsors failed to win support for Rural Action from many of their own regional staff. The shared ownership of the initiative meant that none of the agencies felt responsible for ensuring its success. Moreover, community-led approaches were alien to many network members and more than 60% believed that their lack of expertise in working with local people threatened to jeopardise the success of Rural Action in their counties.

Thirdly, many networks suffered from the lack of a full-time 'champion'. Most network members found it difficult to devote sufficient time to promoting the initiative, and guiding and advising local projects. More than 70% reported that there had been too little promotion and 57% complained of having insufficient time to visit potential projects (Martin *et al.*, 1994). Finally, the level of local interest and take up of grant aided varied considerably between counties.

Partly in response to these findings, several important changes have been made to the arrangements for managing the scheme. In particular, the National Development Team has been restructured to enable it to provide more face to face support and advice to the county networks. Formal written guidelines on the appropriate use of the network grants have also been produced in an effort to remove some of the ambiguity and uncertainty inherent in earlier guidance. Grants have been made available to enable community groups to visit other groups which have tackled similar projects, and to encourage networking at the local level. Nevertheless, networks in several counties remain fragile and highly dependent on the commitment and enthusiasm of key individuals. Moreover, there is still no real sign that the range of community development skills needed to ensure effective operation of the scheme is reflected in recruitment criteria and job descriptions of the staff appointed by the partners at either the national or county levels. As a result, many advisers still tend to treat problems as technical issues rather than long-term community development opportunities. This problem has been exacerbated by the way in which service-level agreements between national agencies and local deliverers focus on tangible 'deliverables' and fail to reflect the time and effort which needs to be devoted to the often largely invisible process of building up the capacity of communities to engage in local action.

The LEADER Initiative

The LEADER (links between actions for the development of the rural economy) initiative launched by the European Commission in 1992 had a number of parallels with Rural Action. Its overall objective was to stimulate local involvement in the design and implementation of projects, which would strengthen the economic and social infrastructure of some of the most needy rural areas in Europe. Unlike the mainstream EU structural fund programmes, which are targeted on deprived rural areas under so called Objectives 1 and 5b and channel funding through the traditional national/regional delivery systems, the LEADER initiative placed particular emphasis on stimulating local development from the bottom-up through new networks and 'grass roots' organisations. This approach was intended to be less bureaucratic and "geared to local requirements and of local origin, making use of available organisational capacity and expertise" (Commission of the European Communities, 1991b).

Local Action Groups (LAGs) performed a similar function to Rural Action networks. They distributed grant-aid derived from the EU's LEADER budget and national programmes and provided training and technical assistance to local people who wished to develop projects (Moseley, 1996). Like the county networks established under Rural Action they enjoyed a substantial degree of autonomy within the overall parameters of a business plan in which they outlined a three year strategy and an indicative range of projects. Like Rural Action, LEADER was intended to promote integrated, multi-sectoral development strategies, which encompassed economic, social and environmental objectives and were tailored to the particular problems and potential of its area. To this end it aimed to provide funding which met the needs of local communities more effectively than traditional grant regimes. In particular there was an intention to allocate 'global grants', which integrated funding from a number of different sectorally based budgets and so encouraged integrated schemes.

A strong emphasis was placed on the importance of disseminating developing innovative approaches to rural development across the EU. To this end a Brussels based team, EIDL, was given the role of establishing and orchestrating a network through which LAGs could exchange ideas and good practice.

The first LEADER initiative (LEADER1) was launched in 1991 and ran for three years. A follow up initiative (LEADER2) was launched in 1994 and is still operational. LEADER1 had a budget of 442 million ECU, which was distributed to 217 rural areas, which between them covered about 16% of the land area of the EU and contained just under 4% of its total population. Most of target areas had low population densities and populations of less than 100,000. Half were suffering from depopulation, most were highly dependent on agriculture and many had relatively high levels of unemployment. The majority of LEADER areas were in Objective 1 regions but all EU member states had some involvement with the initiative (see Table 14.1).

The ex post evaluation of LEADER1 commissioned by the DG6 was completed in 1997-1998. Interestingly, it highlighted a number of very similar conclusions to those drawn from the evaluation of the Rural Action initiative. It found that in many southern European states (particularly Italy, Spain, Portugal and Greece) the initiative was the first ever attempt to promote locally based collective action and to integrate rural development efforts across sectors. It therefore proved difficult to implement, but was also seen as having provided an important demonstration of the value of a 'bottom-up' approach to rural development and of the value of tapping local initiative and potential. It had also acted as a major stimulus for 'horizontal' partnership between local agencies, which had traditionally been very dependent on top-down programmes and funding from their national government ministries.

Most Northern European countries (including Germany, Denmark, Ireland, the Netherlands and the UK) had some prior experience of 'bottom-up' approaches to rural development. National governments in several countries had begun to question the existing paradigm, which stressed unco-ordinated sectoral programmes, controlled by central ministries, and had experimented with domestic programmes designed to encourage local partnership structures to implement locally prepared strategies. LEADER1 provided a further push in this direction. In particular it helped to deepen and broaden local partnerships by drawing in the community, voluntary and private sectors to a greater degree than had been achieved in the past. It also highlighted a number of new opportunities for diversification of the rural economy in the face of reform of the Common Agricultural Policy and had a significant impact on the 'morale' of some communities which benefited directly from grant aid.

Table 14.1. Distribution of local action groups

	LAGs		LAGs
Germany	13	Ireland	17
Belgium	2	Italy	31
Denmark	1	Luxembourg	1
Spain	52	Netherlands	1
France	40	Portugal	20
Greece	26	United Kingdom	13

Note: LAG - Local Action Group

However, the opportunities for trans-national networking and exchange of experience between member states were not valued very highly by local groups and the programme largely failed to promote learning across national borders. By contrast though within many member states approaches and schemes pioneered under LEADER1 were subsequently adopted by mainstream Objective 1, 5b and 6 programmes.

In the UK LEADER1 was seen as having provided a useful encouragement for a more flexible and integrated approach (Bryden *et al.*, 1994; Martin and Geddes, 1998). It enjoyed widespread support among local and regional agencies and encouraged the piloting of new solutions tailored to local needs. It also promoted more integrated strategies across different sectors - particularly SMEs, agriculture and tourism. Regional agencies, such as the Rural Development Commission, Welsh Development Agency and Development Board for Rural Wales, played a key role in encouraging local authorities and other agencies to participate in the initiative and helped to shield local groups from the worst effects of central government and EU bureaucracy. Local Action Groups initiated and supported a number of projects, some of which had important economic impacts and most of which would not have gone ahead in the absence of LEADER1.

However, as with Rural Action, in some areas at least the level of community involvement was small with many schemes drawing on a narrow base of local support and being heavily dependent on the input of professional animateurs (usually the employees of the LAG). Moreover, EU level and central government systems were not sufficiently flexible to allow proper integration of the funding provided by different ministries and directorates.

Conclusions

The importance of both the Rural Action and the LEADER initiatives has been recognised by both UK government and the European Commission. Having been established as three year pilot programmes both schemes have subsequently been extended and allocated increased budgets. LEADER is one of only three existing Community Initiatives which the EU proposes to continue to fund from 2000 onwards. Meanwhile the UK government has applauded the achievements of Rural Action in a range of recent policy statements and, as part of a major new policy initiative, will soon require all local authorities to consult widely with the communities they serve and to involve local people extensively in every area of local policy-making (HMSO, 1998). However, whilst the broad principles embodied by Rural Action and LEADER1 continue to enjoy ever wider support among policy-makers it is important to note that they also continue to pose ideological and practical dilemmas.

In particular 'bottom-up' approaches can be seen as a means of requiring local communities to provide for themselves services which were previously guaranteed by the welfare state. In a post 'tax and spend' era local volunteers run the risk of becoming simply a source of cheap or free labour which compensates or indeed legitimates reduced spending by government agencies. Greater community involvement also has important implications for local democracy since giving local communities a stronger voice and increased involvement in policy-making may well detract from the traditional representative role of local politicians.

More importantly perhaps the evaluations of both Rural Action and LEADER1 suggest that community-led action represents such a significant break with traditional forms of policy-making and local governance that most communities and many practitioners find it very difficult to embrace. It is clear that in most parts of Britain the level of tangible support for local action remains alarmingly small and that the successes achieved by the Rural Action and LEADER1 initiatives depended on the efforts of small numbers of highly motivated and enthusiastic individuals. Even in the most 'animated' communities bottom-up approaches are therefore vulnerable to the loss of a few key people and they may have little to offer areas which lack individuals with the time, ability or inclination to initiate action. Most volunteers are drawn from the ranks of the middle classes and the middle aged whilst young people, the poor and those from ethnic minority communities are far less likely to become involved. There is therefore a need for long-term investment in local capacity building, if the broad commitment to a 'bottom-up' approach is to be translated into sustainable solutions to the problems facing both rural and urban areas. Effective and sustainable local action cannot therefore be implemented 'on the cheap'.

Moreover, there needs to be greater tangible rewards for individuals and for organisations which implement 'bottom-up' approaches. At present the former often find that empowering communities simply undermines their own traditional roles and status, whilst the latter have often entered into partnerships with the community simply because it is a prerequisite for EU and national government assistance.

Finally, the experience of Rural Action and LEADER1 demonstrates the importance of not overestimating the benefits of local action. To date much of what has passed for community involvement has been a re-packaging of relatively conventional small-scale actions, which neither challenge prevailing assumptions nor threaten existing power bases. This has offered important improvements at the margins but not addressed the systemic problems facing many areas. The real tests of the UK government's and the European Commission's commitment to local action will be the extent to which they are willing to provide resources on a long-term basis at local level and whether they are in fact able to embrace local priorities and strategies which are at odds with their own broader agendas.

Notes

1. The British Trust for Conservation Volunteers (BTCV), Royal Society for Nature Conservation (RSNC), Shell Better Britain Campaign (SBBC), National Council for Voluntary Organisations (NCVO) and Action with Communities in Rural England (ACRE).

Chapter 15

Community Readiness: Assumptions, Necessities and Destiny

Bill Ashton

Responsibilities Migrating Down

'Community-based' approaches are emerging as a focus of public policy. For some observers, this might well be signalling a more fundamental change between individuals and their governments (Etzioni, 1993). Others may see it as a practical response as senior government responsibilities, programmes and projects migrate to provincial levels for eventual delivery by communities (Halseth, 1997). Common among many other observers is the remark that sustained actions require both policy (e.g. from senior government) and local actions. The extent of this shift downward was highlighted in an international conference on sustainable communities. People were asked to complete the phrase: 'Community-based ...!' The answers included community-based economic development, social services, senior and public housing, environmental planning, re-integration of young offenders, criminal rehabilitation, strategic planning, organisational development, ... and the list continued (Butler *et al.,* 1995). Both from the top-down and bottom-up, community seems to have gained some increased attention in policy and is a prime forum for action.

What appears to be needed is a better understanding of this emerging public policy thrust and what the keys of success are for a community. This last concern begs other questions about what success is, from whose vantage point, when and how is it measured, and what happens when a community is not successful? Hence this paper on community readiness begins by examining the commonalties of processes that are community-based. This leads to identifying required inputs, expected outputs and the role of planning. For those communities that demonstrate readiness, a self-evaluation framework is suggested. For others that are chronically dependent on government and have limited interest in being on their own, the author explores a more decisive strategy for transition.

An Emerging Policy Response: Community-based Action Approaches

When re-inventing government, communities are increasingly being targeted as the primary vehicle of public policy delivery (Osborne and Gaebler, 1992). It comes as no surprise for some that senior government policies are calling for a community-based approach. In Canada, this downward policy shift is echoed across departments - for example, health, education, human resource development, defence, first nations, justice, solicitor general and housing. Yet there is little evidence of a common theoretical base, a coherent definition or related practice to engender this emerging community-based approach.

In fact there are several prevailing theories, which are used to explain, support and at times justify senior government interventions (Tykkylainen and Neil, 1995). Yet what is certain is that they are amorphous in rationalising a community-based approach policy. In part the difficulty might be from the variety of outcomes from a community-based approach. What is important, argue Tykkylainen and Neil (1995), is a definition of success. Success may mean having more of a balanced growth among the sectors, for instance, or diversification of businesses, or growth in the overall number of businesses, or some scaling down but not a complete loss of firm(s) or a stable unchanged economic status. Others may want to see sustainable development where new business starts are also realising social and environmental objectives. Failures may occur but chronic economic loss also begs a reasonable response. Ironically building bridges among such divergent development processes and interests is more akin to facilitating the design of an approach, than announcing it is part of a policy.

This brief glimpse into the reasons behind the call for a community-based approach needs to be illustrated with some definitions and practices. One way of defining 'community-based' is by examining such activities. Although not comprehensive, the findings of two major reviews of Community Economic development (CED) will be sufficient to draw conclusions: one by Fontan (1993), which is a critical review of Canadian, American and European literature and practices in CED; and the other was completed by Flora *et al.* (1993), which surveyed over 100 development projects and detailed case studies across non-metropolitan USA.

An astute reader would rightly conclude this sample of community approaches has two limitations. The reviewed approaches apply to community experiences, many in rural locations. Thus not all communities were considered. Another bias is that the sample processes focused on economic development generally, leaving a more complete sample and analysis, which includes social and environmental community-based processes, for example, to a later effort.

As noted in Table 15.1, the elements of CED activities are reviewed using an input/output model, which are:

Inputs to CED: What are the necessary start-up ingredients for a community-based approach?

Planning of CED: Is there a particular sequence which practitioners seem to use for understanding how it is done (e.g. through a process that is rigid or flexible, or something in between)?

Table 15.1 Essential elements of community-based approaches

INPUTS TO CED	
a) Organise an initial group b) Establish why change is needed c) Determine decision making process of group	d) Leader gives voice to the 'common good,' moderate risk-taker, support failure, taps into resources - locally and elsewhere e) Use model processes, success stories, build best practices
NEW PARTNERSHIPS EMERGE: As a result of clarifying goals, agreeing on principles and outcomes, an environment is created for partners around the table to make commitments, find ways to inform each other, select a preferred process and timetable, create an initial list of resources, and keep others informed.	

↓

PLANNING CED	
a) Issue-Inventory Establish goals, prepare vision List resources Gather Information, define issues Analyse, select option Publish plan that resolves issues Make public aware	*b) Strategic Development* Determine priority issues and options, and inter-dependencies Select key actions for balanced development Prepare implementation plan Evaluate and obtain feedback
NEW INSIGHTS EMERGE: By knowing what is unique about the area, being aware of the comparative advantages, and having a future focus and an orientation for action, both informal and formal leaders can more easily invite involvement from others.	

↓

OUTCOMES FOR CED	
a) Improve human welfare, greater equity or access to opportunities b) Enlarge the circle of leaders and 'doers' c) Increase jobs, improve employment choices d) Result in a sense of direction	e) Gain support for a shared vision community-wide f) Foster a learning and enterprising culture for all g) Build capacity locally for problem-solving

Source: Adapted from Ashton (1990); Ayres *et al.* (1990); Baker (1994); Chanard (1994); Department of Defence (1993); Garkovich (1989b); Kretzmann and McKnight (1993); Kusmin (1994); Luther and Wall (1986); Muegge and Ross (1993); Nebraska Department of Economic Development (1994), Nelson Community Project (1994); Phillips (1990); Pojul (1994); Pulver (1993); Rural and Small Town Programme (1994); Shaffer (1989); and University of Maine Co-operative Extension (1994).

Outcomes of CED: What is the range of results that are both expected and actual?

Organising the various reviews of CED as an input/output model helps to determine what is perceived to be community-readiness. Observations about readiness from these analyses, even to a casual observer, include: preparedness for dealing with complexity; knowing when to lead and when to manage; benefiting from an enabling attitude; managing changes; and viewing CED from different perspectives. Each of these observations will be explained, before suggesting a framework for determining community-readiness for CED.

Preparedness for Dealing with Complexity

Implementation of strategy is a complex task. While many processes discussed leadership, none of them identified the skill set that is needed to take action, to risk and partner through implementation (e.g. given a work or action plan, coupled with the need to gain access to capital and to be innovative, while matching the projects with existing and available resources, and interests of others). Indeed, with more than 17 years in this field, the single largest, most complex challenge, which remains, is implementation - making change happen.

Knowing when to Lead and when to Manage

Little attention was paid to those who should keep on with the strategic efforts (e.g. sharing the vision, gaining more support, finding new partners, completing environmental scans, scenario-building and inventorying resources). The point is that there is a difference between managing - doing things right after implementation - and leading - implementing the right actions to move towards the vision (Rost, 1991). Once implementation begins, three or more groups may be required to keep the momentum going - some leading the strategic efforts, others dealing with implementation and others managing to extend invitations and finding ways for getting other people involved in and beyond the local community.

Benefiting from an Enabling Attitude

Often an agency was the prime advocate and means of liaison with the community group. Much work is still needed to foster an enabler/contributor role, rather than the more traditional provider/dependent approach. These two fundamental roles are discerned from these cases, as seen in Table 15.2. The outcomes from an enabling agency include, for example, establishing performance targets, trusting that solutions are found through local projects and believing that local champions are important for success. All of these are necessary for a leader to facilitate action.

Managing Changes

The reviewed CED processes, in varying degrees, all pursue some aspect of problem-solving for managing change. For some it means the traditional top-down approach where the government or an agency is the 'provider' (e.g. there are problems to solve, the community is a consumer and so on). Yet, there is a second view, which is exemplified as a community-based, bottom-up approach. Here the initiating agency or government's role is to enable the use of the community's assets to focus on opportunities. This requires an outside agency to value the importance of a community, to nurture relationships and to support a community in becoming more of a producer of its own wealth. Some of the distinctions between top-down and bottom-up processes that were found in these CED projects are noted in Table 15.3.

Table 15.2. Role of government

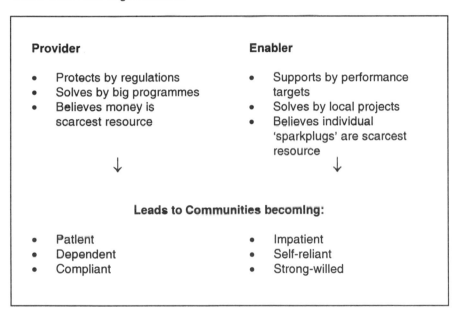

Provider	Enabler
• Protects by regulations • Solves by big programmes • Believes money is scarcest resource	• Supports by performance targets • Solves by local projects • Believes individual 'sparkplugs' are scarcest resource
↓	↓
Leads to Communities becoming:	
• Patient • Dependent • Compliant	• Impatient • Self-reliant • Strong-willed

Source: Adapted from Phillips and Williams, 1991; Sears *et al.,* 1992; Sears and Reid, 1995; Voth, 1989.

Viewing CED from Different Perspectives

From this review of community-based approaches, there are several ways of looking at economic development in a community. The traditional economic development model tends to be top-down (e.g. led by government involving large investments), while bottom-up CED is led by community leaders and local organisations. Notwithstanding this apparent difference, well-intended governments and agencies still use the term CED, but remain top-down in their approach. To others, CED indicates there is a local organisation involved and some consultations will be initiated for 'grass-root' input into a larger process beyond the community. For most advocates this falls short of the desired local involvement and local control of a CED effort.

Table 15.3. Two approaches to managing change

Traditional 'Top-down'	Community-based 'Bottom-up'
Needs-based.	Asset-based.
Problem-oriented.	Opportunity-oriented.
Community is externally focused.	Community is internally focused.
Driven by needs assessment.	Driven by community vision, leadership, and commitment.
Community is a consumer of services.	Community is a producer of services.
Only experts can provide real help.	Self-help. Communities are expected to do it for themselves.
Aim is to build bigger and better community services.	Aim is to relinquish control of services to communities whenever feasible.
Government protects by regulations and controls by policy.	Government supports and encourages by performance targets, criteria, and benchmarks.
Big programmes solve problems.	Local projects solve problems.
Citizens consulted in policy-making.	Local citizens invite others to form policy.
Public servant is regulator, service provider and expert.	Public servant is facilitator, coach and supporter.
Community groups, individuals, businesses, and local Government work in isolation and compete with each other.	Community groups, individuals, businesses and local government collaborate to build partnerships and alliances.
Rational, comprehensive planning, limited action.	Integrates, blends planning, pilots and actions.
Learned	Learner

Source: Adapted from Kretzmann and McKnight, 1993; Phillips and Williams, 1991; Osborne and Gaebler, 1992; Muegge and Ross, 1993

Outcomes from CED initiatives are often neither immediate nor easily measured. From the processes reviewed and from the conclusions reached by Flora *et al.* (1993), efforts solely evaluated on the basis of the number of jobs created/saved or the amount of income generated in the community are not a viable strategy for non-metropolitan communities.

Zekeri (1994) had a similar finding from an examination of 120 rural areas using five categories of variables including: promotion to attract business and industry, financial assistance to firms, creation or expansion of recreation activities, availability of human services and government programmes. He also considered seven community characteristics, including the accessibility of the location. His analysis made two points: it indicated that practitioners and policy-makers should focus attention on an organised intentional effort; and there are differences (e.g. structural and functional) in communities and these should be acknowledged in their economic development strategies. These five aspects characterise the ingredients, activities and outcomes of community-based approaches. The next section provides a framework for readiness.

A Framework for Community Readiness

The apparent beginning point for readiness is at the 'getting started' stage of a community-based approach. Conceptually, getting started lies between the time when citizens are concerned and the time when they want to take collective action. Having studied, taught and practised CED in North America and overseas, Bryant (1994) also concludes that attention is best placed early in the 'input' stage.

A self-evaluation framework of community readiness is one possible response, as noted in Table 15.4. It is applicable for the beginning stage of a CED activity. The three primary concerns are described with questions. Answering them helps a community group to understand anecdotally if they and others in the community have a profile or portrait of readiness. These concerns help assess the local history for community involvement, gauge the degree of co-operation among groups in a community and examine partnerships for actions. Thus, from the beginning and almost irrespective of the CED process to be used, the community must see itself taking charge and achieving results. What this framework of 'readiness' includes are the key considerations for success in all three stages of community-based approaches - inputs, planning and outcomes.

Community Participation

Nearly a dozen questions probe the three basic aspects of participation in a community: who would be most likely to be interested in initiating a change in the community; how would the community define itself - at least geographically; and are the key individuals available who are most likely to form a new partnership in the community? These questions are not intended to garner a yes or no response. They challenge one to think about who, from the community, might be involved and the nature of their contribution.

Community Co-operation

Stepping away from the individuals, the five questions consider the abilities of organisations to renew their membership. This can be achieved by working with other groups, involving volunteers and also by making information easily available.

Community Partnerships

These following questions examine how organisations work together. What evidence is there of organisations forming partnerships that last? What role have community leaders played in rallying broad support from the community? And to what extent have organisations and clubs systematically initiated and completed joint projects? Answers to such questions will suggest if the community has an inherent propensity to bring about a future based on collaboration.

Table 15.4. Framework for community readiness: getting started

---

1. COMMUNITY PARTICIPATION   ↓

---

a) Principal Stakeholder Organisations
- Who are they?
- What is their interest(s) in sustaining the community?
- What is their track record of contributions (broadly defined)?
- What is their potential for greater involvement?

b) Your Community Defined
- What is the area or geographical extent?
- Are all interests represented or are there sufficient interests to define why change is needed?
- Are the aspects critical to sustaining your community defined?

c) Individual Participation
- Are the partners able to speak from experience when representing the interests in the community or are there model processes, success stories for reference?
- Are the partners involved in several organisations already? - busy people get things done.
- Have the partners been involved with informing the residents and asking for their involvement?
- Can the partners rally support from their organisations?

---

2. COMMUNITY CO-OPERATION   ↓

---

a) Organisations Working with Others
- What examples are there of organisations in your community co-operating in projects for several years?

b) Replacing Volunteers
- Is there a planned replacement of volunteers in your organisation?

c) Information Availability
- Do residents have easy access to plans and information?
- Are community forums held, and which organisations regularly partake and contribute to them?
- Do you regularly communicate with people who are not part of an organisation?

---

3. COMMUNITY PARTNERSHIPS   ↓

---

a) Organisations as Partners
- Do organisations (elected and volunteer) have track records of entering into long term partnerships with others in your community?
- Is there an organisation for community (economic) development with a track record of building lasting partnerships with others and who have a role to play in sustaining your community?

Table 15.4 *contd*

b) Community Leaders
- Have there been periodic events like forums, workshops and celebrations for all segments of your community to participate in?
- Are people encouraged to come forward and participate in collective decision making?
- Have the underprivileged of the community been invited or been assisted to be involved?

c) Joint Actions
- Do boards periodically (annually) get together to plan joint activities?
- Is your organisation willing to invest in co-ordination to reduce duplication of efforts in your community?
- Are the boards willing to share the responsibility, the credit or the blame in an equitable way?
- Is there a commitment to spend the time to communicate effectively with their partners?

Source: Adapted from Bryant (1994); Rural and Small Town Programme (1994); Blakely (1994)

Testing the self-directed instrument, as noted in Table 15.4, and refining it would seem to be an appropriate next activity. Such work by the senior government (i.e. federal, provincial) could provide insights into expected utilisation rates of a programme requiring a community-based approach - data usually sought in making a business case for a programme.

Certainly other questions are left unanswered from Table 15.4. While there may be evidence of community participation, co-operation and partnerships in a community, what else is needed? Is there an indication of willingness of the community and an organisation to take on another set of activities resulting from another community-based approach? What are the local abilities to handle the start-up, the planning and the implementation amid all the other demands in the community? In other words, are there times in a community's life that they reach a limit to taking on another community project?

It is easy to imagine that communities probably reach thresholds with their abilities and willingness to manage their current and future situations. Although a threshold for community involvement is conceptually possible, yet it has not been possible in a practical way. In part, one would expect a threshold to differ over time and among communities. But what is evident is the limited expression by senior governments and the communities themselves to begin a discussion about the most strategic use of resources in communities.

Readiness should require communities to be much clearer about taking on additional activities. In other words, communities may be stating a qualified 'yes' to conditional funding. Currently the 'down-loading of services' has assumed that all communities are able and willing. Knowing their own thresholds might leave the community negotiating room to adjust priorities in getting to a yes, while keeping focused on their own shared vision. The conditions of a qualified 'yes' would hopefully be sufficient to ensure the outside agency is a partner with the community.

From an outside agency's perspective, the above discussion might well signal an increasingly competitive community market place. As communities are being asked to take on more, their abilities and willingness will diminish with each consecutive wave of 'community-based' programmes. Thus the outside agency's programmes will be competing for community attention and participation. This may, as some put it, leave the next community-based effort short changed. The valuable and limited resources of volunteers will have already been committed.

Responding to Chronically 'Unready' Communities

Simplistically, two types of communities emerge from the discussion:

Ready Communities. Those communities where improving their odds - with participation, co-ordination and partnerships - is part of their culture, and which therefore demonstrate their abilities to stem a decline and remain economically viable and a pleasant place;

Unready Communities. Some rural areas are likely to continue to decline and lose economic viability. As Walzer and Ching (1994) conclude after assessing hundreds of rural and small towns in the state of Illinois, "...unless they [community leaders] take charge of their communities' futures, many will, in an economic sense, disappear."

This paper, consistent with much of the literature, has focused on communities with a 'personality' for growth or development and sheer staying power. Hence, they have a self-proclaimed readiness for CED, albeit to varying degrees. What often goes unsaid is what about those communities that are unready? Those communities that are consistently teetering on 'bankruptcy' - socially with council after council resigning, economically when senior government grants are double or triple the revenues from local taxes, or environmentally where a mine is killing residents. This section, therefore provides a discussion about a response for those chronically 'unready'.

What about communities which are declining and disappearing? Rarely do communities completely disappear; something usually remains. Often they become part of a larger community. Certainly their abilities to provide services will diminish. Yet, for some communities heading towards such a fate, just mention of their situation by a senior government officer might be sufficient to put them back into a survival mode. However, after several 'resuscitations' and no organised voice, both Luther and Wall (1986) argue that it might be hard to ignore the silence that suggests there is a satisfaction with decline. In medical terms, a watershed event from living to dying takes place. Palliative care or terminal care are terms used to describe the treatment of dying with dignity. Is there sufficient reason to assume such a process might be of use for communities?

Meanwhile, senior governments, with a political right-of-centre bent, have a growing interest in investing public moneys where risks are modest and returns mean making a contribution to recognised markets. However, the risk is high. Any government or expert taking up this palliative care notion would soon be labelled,

'Dr Doom' or 'Mr Death' or the 'Department of Decline'. But behind the names and rhetoric, is there a genuine need for some discussion?

After many attempts and likely years of decline, questions remain about which communities will survive and which should receive additional injections of public money. Unlike life itself, death is not the only option for communities. There are many communities, which have 'seen death knocking at the door' and turned the situation around. Like any business, successful communities also claim their strength is the people. However, if only a few residents remain, perhaps a strategy dealing with a transition is not so far fetched. This could be a starting point to an important topic like the community's new future.

Bridges (1991) notes, new beginnings usually require something to end. From the community's perspective, they may well appreciate clarifying their specific contribution to the province and nation! This in turn would better equip the community leaders to set realistic development strategies, while arming themselves with key arguments to attract investors.

In reviewing over 900 articles, books and abstracts and having met with medical directors and expert geriatricians, Keary *et al.* (1994) observed five common aspects of palliative treatment. These are noted in Table 15.5 along with an analogous community situation and possible strategy. Terminal care treatment is unique for each patient. The intention is to preserve their autonomy, as it should be for each community. However, this may require the community to re-define itself with new boundaries. For example, a failing village in New Brunswick has options - annexation of the rural areas, dis-incorporation into a rural area overseen by the provincial government or becoming a rural community.

The debate about dying with dignity has parallels for communities in chronic decline. Palliative care treatment involves documenting early on the patient's wishes, including directives for administering life-sustaining interventions. Throughout, professional assistance is essential and ongoing support takes many different forms, for the patient, the family and others. Likewise it would be important to identify the impacts which 'dis-incorporation' would have on other communities and the provincial government. Undoubtedly with unincorporation, there will be a shift of responsibility to the province - financially and politically. However, what is important is that the residents are informed and have lead roles in their own transformation as a community.

But what is the entry point to privately or publicly discuss the eventual demise of a community? For most communities in chronic decline, there is no way for a senior government, let alone their own local leadership, to discuss such an option. If there was some concern expressed about declining communities, would that itself be a sufficient motivation for survival? Further work is required to understand the dynamics of consistent decline before strategies are imposed. What is clear is that senior governments will want to make continued adjustments to communities and their financial commitments as they tackle the larger and much longer term issue of debt reduction. Such a longer term perspective is consistent with bringing about changes among communities. However, will this situation create sufficient interest for a debate about readiness and its unwanted cousin, community decline, by either the senior governments or the communities themselves?

Summary and Conclusions

In a period where change is rapid and complex, communities as well as countries face many challenges. For rural communities, the challenges seem to be even more daunting - with demographic shifts working against revitalisation (Bollman *et al.,* 1992; Government of Canada, 1995), with capital being substituted for labour (Bollman *et al.*, 1992), with local communities joining together for economic opportunities (Annis *et al.,* 1994; Holtkamp *et al.,* 1997) and with equity issues among the communities wedging them apart (Ashton and Lightbody, 1990; Robinson, 1995). The emerging public policy is one focused on community-based actions.

There remain many assumptions and unknowns about a widespread shift to a community-based approach. A community-based approach to service delivery, when examined in the context of rural areas, requires some reconsideration for several reasons. Even the basics of readiness, which are thought necessary for a community-based approach are likely to vary widely. Increasingly, the apparent down-loading of programme and service delivery will come up against barriers of unwillingness at the local level. In addition, there is a large variation in knowledge and skills required to carry out successfully a long-term community-based commitment. These very factors may help communities realise that a new way of doing business is needed. Both communities and governments will have to choose strategically which services they can deliver, and when, and under what conditions. Likewise the initiators, senior governments and private sector, are also going to have to change from provider to enabler. As a result, both communities and initiators have an interest in community 'readiness.'

When 'readiness' means preparing and implementing CED, this paper concludes the way to improve the odds is to focus on the 'getting started' stage. Specifically the focus is on community involvement, co-operation and partnerships, all related to the demonstrable willingness and proven abilities of people. The focus on people is most appropriate. People build communities and they collectively contribute to a prosperous province and nation.

This paper also raised two other concerns. The first is with 'how.' What are the process(es) and skills needed to carry out community-based approaches? The literature indicates an evident lack of both a generally accepted process for a community-based approach and no consensus on its evaluation. One first step might be an investigation into how systematically to uncover aspects that are predictable and useful in advancing community-based approaches - both the practice and the theory. Does the concept of using an International Standards Organisation (ISO) approach lend itself to this challenge? Nevertheless, having predictable results with a community-based approach will continue to be a linchpin to the nation's public policy future.

A second concern raised in this paper is the antithesis of 'readiness' which is the demise of communities. It is recognised that communities do not die; they become part of a larger community. However, there is no certainty or predictability about how a community evolves into a rural environment. Thus, parallels are drawn from the medical field for insights about palliative care. Admittedly, more work is required to explore the public responsibilities associated

Table 15.5. Palliative care strategies for communities

Care Strategy and Medical Description	Community Description	Response to Community (negative scenario)
Pain Control		
Mild, general (aspirin) to sharp and intense (morphine), including undetected fear ↓	Continued high unemployment, loss of jobs, high cost of living	Identify and manage current problems. Council adopts 3 year recovery plan, supported by provincial government.
Relieve Shortness of Breath		
Ameliorate if possible, provide on-going relief if necessary ↓	Natural disaster, rapid loss - bust, widening financial situation, major infrastructure set back	Focus resources on problems related to survival. Village council assist, trustees appointed if council dissolves.
Psycho-Social Support		
More than basic continued involvement, apply resources for counselling, emotional, spiritual support for patient, family and others ↓	Low participation, hope-filled future only a flicker, civic responsibilities neglected, leadership infighting, diminishing volunteer sector	Keep people informed. Services limited, no investment in capital works. Help people re-locate.
Hygiene		
Physician conveys importance of hygiene, key for patient comfort, dignity and well-being ↓	Abandoned or unkempt buildings, deterioration, consistent neglect for services and facilities, high dependence on government	Options for community's future discussed publicly. Choices made. Few services public, co-operative or private.
Symptom Control		
Attention to other symptoms causing discomfort, each symptom actively evaluated, often followed with intervention	Unable to offer services locally, failures impairing future opportunities, growing poor and disenfranchised, high dependence, obvious neglect, unable to respond	4 years later. Community annexes more area or dis-incorporates (larger provincial government role). Local government structure transition over. A new beginning for those who remain.

Source: Adapted from the five step and medical description T. Keary *et al.* (1994); Humphery (1992); Hoffman and Dupont (1992)

with communities in chronic decline. Nonetheless, the call is made for both communities and provincial governments to fill this gap, which has to date been an unspoken policy frontier. If this was a topic in the policy arena, the senior government could see it as a plank in their investment strategy, while opposition will see it as a strike against equal opportunity.

There are several responses to the emerging community-based approach used by an increasing number of senior governments. This paper has deliberately focused discussion on identifying the keys to success. The assumptions and necessities of readiness for a community-based approach means putting the community first, in a lead role. Moreover, communities in this paper are seen as the preferred point for inventing the future with creativity and innovation, and through experimentation. Such a destiny is possible when communities are seen as the engines for wealth creation - in terms of economic growth and development, enhancing human capital and managing environmental capital.

Notes

1. A special thanks to Alain Bryar for his comments and suggestions on the final version of this chapter.

Chapter 16

Rural Economic Diversity: Adaptive Strategies for Minnesota's Boreal Forest Region

Catherine Lockwood

Introduction

Diversification and multiple occupations are key elements for economic survival in Minnesota's Boreal Forest Region (Fig. 16.1). Adaptive strategies based upon clever use of area resources by local residents result in a distinct economic survival pattern. A common belief for most rural areas is that agriculture provides the basis for economic stability, however, for people living in marginal areas, such as Northern Minnesota, this is not true. Instead, these people have devised alternative tactics for identifying resources, using the land, and making a living. Similar to other rural areas that claim economic dependency upon marginal lands, Northern Minnesota has an inhospitable climate, low population with a scattered settlement pattern and limited resources. Cold weather, a short growing season, poor soils and bogs, while adding dimension to survival, underscore the reality of a marginal area that must depend more on the ingenuity of its people than its resource base for economic stability. What do people do to survive in a marginal region? Since Minnesota's Boreal Forest Region is a marginal area with limited opportunities, several questions are addressed: who are these people, what do they do and how do they survive?

The research of Northern Minnesotans presented here is based on seventeen years of observation and numerous interviews. By investigating historical and current land use practices and job choices, four groups of economic activity have been identified - seasonal, year-round, high-risk, and nostalgia - that explains the complexity of alternative economic subsistence strategies for a region with minimal resources and a sparse settlement distribution. These categories divide activities by type including subsistence, labour, extractive, exploitative, cottage and professional endeavours. Classifying employment explains how people cope within a remote area and why people choose to move into or remain in a region with limited economic opportunities. Three groups of people have also been identified by length of residency and by personality traits, which helps explain individual economic survival choices within a remote, marginal area.

Fig. 16.1. Minnesota's Boreal Forest Region

Settlement and Land Use Succession

A misplaced supposition about rural areas is that most people make a living by farming. This is not so in Minnesota's Boreal Forest Region. Settlement was a result of overlapping perceived opportunities. Early economic development depended upon resource extraction for profit including mining of taconite and cutting of timber. The Great Pine Forests seemed an endless source of revenue and products. Logging companies encouraged settlement through aggressive advertising and joint ventures with railroad companies (White, 1980, pp. 106-107). The Northern Pacific railroad brought people and supplies into and hauled lumber out of the region. Land was cheap, available and accessible. Promoting agriculture and selling farmland was a way of converting land where trees had been felled into usable income. Some people moved into the area with European farming techniques, but quickly adapted if they wished to survive and succeed, especially in a region with a cold climate and poor soils that limit agriculture. Northern Minnesota tried to compete with the rest of the United States by supporting itself with agriculture. Despite incentives offered by the Minnesota Department of Agriculture during the 1920s and 1930s (Committee on Land Utilisation, 1934), those who lived here then and those that live within the region now recognise that a single endeavour such as farming is not a sustainable venture. Although farming declined after 1930[1], many agricultural characteristics remain as a part of the current landscape and land use pattern. Agricultural icons give the impression of 'farm country,' but the region does not function in an agricultural capacity. Most land either is not suitable for agriculture or is pre-empted by other land uses. Over 60% of the land is in public use, mostly state and national forests, wetlands, and wildlife preserves. The remaining land is a mix of peat bogs, swamps, partially wooded areas and minimal open space (Land Management Information Center, 1995a, 1995b). While people still engage in agricultural types of activity, most have other means of support.

Between 1940 and 1960 farming continued to decline while mining and logging industries remained stable. A sluggish economy and static population growth resulted in little regional change. Between 1960 and 1980, however, economic development was on the rise and formal business began to diversify (Anding *et al.*, 1990, p. 23). Recreation was recognised as a resource in the 1930s, but did not emerge as a significant part of the economy until the 1970s with the exploitation of natural features and cold weather for winter sports. During the 1980s, state rural planners continued promoting the region's outdoor recreational potential. The recreation amenity-based economy remains constant for the 1990s because state planning agencies continue promotion of winter activities. A common saying among local residents is that Minnesota has four seasons-June, July, August, and Winter.[2] The climate and physical landscape are key factors for personal and economic survival within Northern Minnesota.

Understanding the Physical, Cultural and Economic Setting

Minnesota's Boreal Forest Region covers more than 72,500 km^2 with only 11% of the state's population living within the 15 county study area (Fig. 16.1). A dominant land use map shows that while pockets of land suitable for farming exist within the region, most land is forest combined with water, mining or some pasture (Fig. 16.2). Second and third growth mixed forests cover over 66% of the total land for these 15 counties. Open water covers another 18% with the remaining 14% in roads, residential, and cultivated land.[3] Glacial ridges and

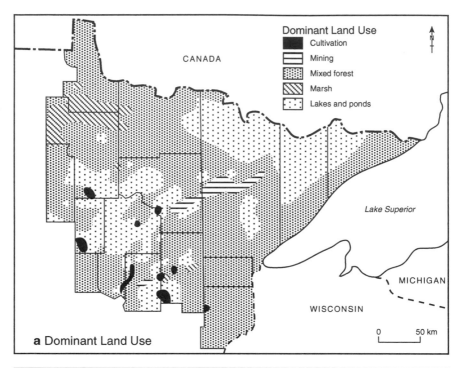

a Dominant Land Use

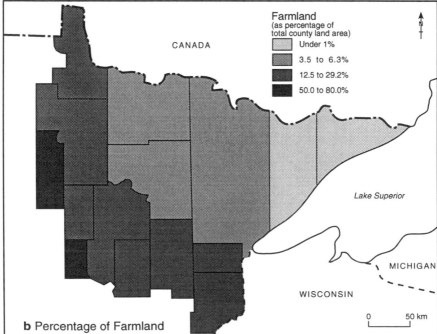

b Percentage of Farmland

Fig. 16.2. Land Use and Farmland in Minnesota's Boreal Forest Region

terminal moraines sharply divide non-farm forested Minnesota from southern oak-savannah Minnesota and from rich alluvial farm and prairie lands to the west.

General Land Use

The 1992 Census of Agriculture helped explain land use and agriculture activity. The map shows a simple progression of farm land distribution declining from west to east (Fig. 16.2). This can be deceiving, however, because only 11% of the total area for the 15 counties is land on farms. Total land on farms for the 15 counties is less than 4% of all land on farms in Minnesota. The rest of Minnesota has intense agriculture with most counties having between 50 to 80% of land on farms (US Department of Commerce, 1994).

While these categories show the extent of farming, they also show physical limitations and dominant economic zones. For example, counties with 12.5 to 29.2% farmland form a transition zone from alluvial soils of the Red River Valley to acidic soils common for pine forests. Here, the local economy is a blend of farming, logging, and service and professional occupations. The farming, forestry or fishing industries employ less than 6% of the Region's work force age 16 and over. Logging remains important to the region's economy, especially for central portions of the Boreal Forest Region. More than 200 logging companies and more than 140 wood products manufacturers compete in the upper half of this mixed land use area.[4] The logging industry is a capital intense, yet low labour market because most companies have replaced labour intense jobs with streamlined operations, a result of improved technology. For example, one person and one skidder can do the work of a three-to-four person crew.

Taconite mines and tailing ponds overshadow land use in eastern Northern Minnesota. The local economy depends upon resource extraction (mostly mining), construction (new housing for recreation or retirement and winterising of cabins) or recreation and tourism (family-owned lake resorts and winter sports like snowmobiling and ice fishing). With less than 1% of the land on farms, farming is virtually non-existent to the north and north-east. Local residents call the region a 'farm area gone bust'. What little farming there is within the region is innovative and varied. Those who farm add to their income with off-farm jobs, a working spouse or retirement moneys.

The People

The character of the people affects lifestyle and economic choice. Two words encapsulate common qualities of Northern Minnesotans: independence and perseverance, and in turn, these traits strongly influence what people do to support themselves. People resist outside pressure and change. Mamie Pheifer has lived in the region for more than 25 years. She chose Northern Minnesota so that she could do as she pleased answering to no one but herself. Mamie's beliefs are similar to those of New England where she was raised. She meshes with the region's population applying the attitude that she can make do and get by (Pheifer, 1995, personal communication). Before her arrival in Minnesota, Mamie and her husband John moved across the country searching for more than a job. They wanted a lower cost of living, privacy, trees and a safe environment for their children. Mamie balances quality of life with less money and accepts jobs that others did not want so that she can remain in Northern Minnesota.

Besides recognising an alternative economic strategy for the region, Northern Minnesota has distinct groups of people with common traits, length of residency and livelihood. Life-long residents were born, raised and remain within the region. Life-long residents, in some respect, are realists. They are durable people, who adapt to a severe climate and restricted economic choices with stoic acceptance. These people are content and seldom consider living anywhere else. The second group, returning locals, were living elsewhere with the hopes of a better job, attending school or serving in the armed forces. Now, for varied reasons, often as simple as the need to care for elderly parents, they are returning to family and home. The third group, newcomers, includes two types of people: recent arrivals, residents of 15 to 20 years; and new arrivals, those with less than five years residence within the region. Newcomers want privacy, lower crime rates and a healthy place to raise their children, yet they bring city ideas with them that often conflict with the traditional, conservative values prevalent throughout Northern Minnesota. Newcomers often have difficulty accepting the lack of cultural events such as opera and baseball games. One recent arrival believed that he was culturally suffocating, and although he would like to leave the region, remains because of his job as a local high school teacher. A third group of people, who have been resident for 6 to 15 years, exists by default. They moved to the region with enough money to invest in a 16.2 ha section and remain because they do not have enough money to leave.

Life-long residents are cautious of newcomers. Amy Goslin's parents homesteaded in Itasca County in the late 1890s. Mrs Goslin was emphatic about the distinction between residents and newcomers. If a person was not born in the area, then they are 'new' to the area.[5] Her son-in-law's family moved to Northern Minnesota in 1928, but to Amy, they are newcomers: "I've gone out a couple times and worked, but I always came back. I raised a family here. But not many [old-timers are] left. There's nobody-just my daughter's in-laws, they came in the twenties or thirties so they're not old timers either" (Goslin, 1992, personal communication).

The Economic Structure

The Census data for official employment of the Region's residents provides an understanding of its economic structure. Over 60% of the workforce were employed, with two-thirds of those in employment working more than forty weeks per year. Forty per cent of the potential workforce was not employed and only one-third of the employed worked more than 40 weeks a year. Because of this underemployment, 16.0% of the population survive on an income below the poverty level. The Boreal Forest Region had an unemployment rate of 9.1% compared with the State unemployment rate of 5.1%. The majority of rural residents receive a wage or salary, yet they also receive some form of supplementary income such as social security, retirement, self-employment or public assistance (US Department of Commerce, 1992a, Tables 6 and 9).

By computing Location Quotients (LQ)[6] for the 15 counties, the relative importance of Census defined occupations and industry within the Boreal Forest Region are revealed. Only the service economy is more important to the region than to the state as a whole. Service tended to be 8 to 14 times more important in the study area. Administrative jobs are least important. Mining is a dominant industry for Itasca, St. Louis and Lake Counties, the location of the Mesabi Iron Range. Farming is insignificant except for Clearwater, Pine and Wadena Counties, which are transition counties between the Boreal Forest Region and more prosperous farming areas to the west and south. Land use practices for Clearwater County conform to agriculture practices of the Red River Valley, while Pine and Wadena County's land use patterns broadly stretch the upper limits of the Corn Belt. Beltrami and Itasca Counties depend heavily on wholesale trade and on services. Beltrami County also relies on education as a basic industry because of Bemidji State University located within the county (US Department of Commerce, 1992b).

The US Census, however, does not adequately explain the nature of work or choice of occupation by residents within Minnesota's Boreal Forest Region. Two questions need answering: who are these people and what exactly do they do to survive economically within the region?

Rural Economic Strategies

Economic strategies are as mixed and as varied as the people who live in Northern Minnesota. Rural residents adjust to harsh physical, economic and social situations as an accepted condition for living in Minnesota's Boreal Forest. Their ability to adapt is a reflection of their social and cultural background. These are fiercely independent, family oriented, conservative individuals.

Few people within the region have a single job. Logging in fall and winter, farming in spring and summer, with odd jobs throughout the year were common when Northern Minnesota was first settled. The trend continues with multiple occupations still a viable means of earning a living. Often, local residents with full-time or permanent part-time jobs, supplement their income by other mechanisms, such as bartering and exchanging favours, which they do not claim as income.

These responsible, tenacious people view work as their entertainment and as their leisure activity. Jane Carlstrom, a registered nurse and her husband Frank, a retired military officer and civil engineer, choose to farm: "The farm is our lifestyle, which we support with outside work. This is our pleasure. We don't ski; we farm. We don't golf; we farm" (Carlstrom, 1993, personal communication).

Classification of Economic Strategies

Adaptive strategies based upon individual ethics and clever use of area resources result in a distinct economic survival matrix. Table 16.1 shows four dominant categories of economic strategy. Each category has selective traits that result in an individualistic lifestyle choice by the local residents. The Seasonal and Year-round categories that represent ways in which people use available resources are divided into type of industry including subsistence, labour, extractive, exploitative, cottage and professional endeavours. The term industry designates the method for living in a harsh, remote area and should not be confused with industry as a formal business. Although High-risk ventures and Nostalgia lifestyle are subsets of Seasonal and Year-round, the use of non-traditional economic tactics and idyllic lifestyle choices warrant separate categories.

Table 16.1. Northern Minnesota's rural economic strategies, selected activities and traits

Category	Type of Industry	Activity	Significant Traits
Seasonal	Subsistence and Diverse	Wild rice; Maple sugar; Vegetables; Berries; Small game	A hunting and gathering lifestyle; subject to land access, available resource and experience
	Extractive	Logging; Mining	Occupation divided by summer, fall and winter
	Labour	Trucking; Carpentry/ construction	Gender and age specific
	Exploitative	Resorts; Recreation; Tourism	Family business; employs college and high school students
	Cottage	Crafts; Wreaths; Fish ties	Independent suppliers; gender specific; low income
Year-round	Seasonal - intense	Large dairy or sow operations; U-pick-berry farms	Main income from large scale diversified farming, supplemented by large scale seasonal subsistence activities
	Seasonal - diverse	Wild rice; Maple sugar; Berries; Vegetables; Small game	Combination of seasonal enterprises that create a pattern of cyclical subsistence self-employment
	Extractive	Pulp and paper operations; Chip board and fuel pellets	Wood products processing as a continuation of logging
	Professional, labour and service	Wholesale and retail trade; Government; Education	Full- and part-time employment based on demand and need; also uses seasonal activities
High-risk	Exotic	Non-traditional livestock; Ostrich and emu	Large investment – quick and profitable return
Nostalgia	Historical farming	Seasonal gathering; idealised farming	Supported by other income; lifestyle choice

The lists are not exhaustive and show general occupation strategies with examples to illustrate the classification. In some instances, activities overlap categories because of resource use and personal preferences. For example, wood production is both a seasonal and a year-round activity, but not for the same individuals. Extraction of timber begins in late fall and continues through early spring when the first thaw results in highway weight restrictions. Logging companies ship bolts of logs hauled from the forests during winter months to regional plants for processing into chip board, fuel pellets or paper products, which keep production plants in business year-round.

Exploitative activity stems from the recognition that by redirecting outside perceptions about a severe climate, by promoting a recreation landscape and by fostering an idyllic lifestyle, rural is a resource. The lifestyle is rural with subsistence activities as leisure pursuits. The region is evolving from felled forest lands with extensive regrowth and marginal farms, to a rural leisure landscape. Recreation and tourism are emerging as vital businesses for Northern Minnesota. Minnesota is the Land of a Thousand Lakes and the Boreal Forest Region has more than its share. The International Boundary Waters recreational area is a major destination point for individuals who want a wilderness vacation experience. Many resorts are family operations, however, often either one or both spouses have other jobs such as teaching to support themselves during the off-seasons. Northern Minnesota follows the national tradition of a three-month summer season beginning the last weekend of May to the first weekend of September. Even under the best and most profitable conditions, summer tourism must be augmented by other income. Currently, other seasonal leisure choices such as winter sports and hunting seasons are creating a year-round recreation economy. Tourism offers additional summer income for area residents with lakeshore homes or a cabin or two. They rent these units during the summer or for various hunting seasons. Some local families camp at a designated relative's house while they rent their houses for the opening weekends of high demand sports like fishing, duck and deer seasons. High school and college students provide minimum wage labour for the summer recreation season.

Seasonal and Year-round Activities

Most people have some type of day-job, either full-time or part-time, but raise extra cash with traditional and creative side occupations. At home jobs such as sewing and childcare are common for women, while small engine repair and cabinet making are common occupations for men. Local residents have part-time businesses such as life-insurance companies or in-home beauty salons because they recognise that a single endeavour such as farming is not a sustainable venture. Jackie Olsen, a retired schoolteacher, has lived in the region her entire life. She and her husband bred and raised beef cattle on 315 ha of wooded and rocky land. Without additional income, the Olsens would not have survived 30 years of marginal farming.

"My husband had a little insurance company, he sold insurance, and drove [a] school bus, and logged, and worked on the highway. I used to complain that farming was a terribly expensive hobby because you have to do all these other things to buy machinery, which was expensive. It wasn't always easy. [But] we never once thought of leaving; never once." (1992)

Some local residents combine food-stuff production, indigenous plant gathering and seasonal occupations into a year-round-diverse, self-employment pattern. They supply local grocery stores with surplus garden vegetables, while local markets provide an opportunity for peddling home-processed foods such as honey, jams, baked goods and eggs. Seasonal cash crops, specifically wild rice, collecting and processing maple sap and wild berry gathering, supplement their incomes.

Wreath production is a lucrative seasonal enterprise. Beginning in October, buyers advertise in local papers for palm boughs, paying up to a nickel for 0.5 kg. The season is over by the first of December. Tying fish lures is also seasonal work, mostly during the winter months and mostly by women.

Cottage industries, especially crafts, though classified as seasonal, can be year-round activities with seasonal peak periods of production. The region's largest cottage industry, Lady Slipper Designs,[7] has about 25 participants who regularly work on a per-piece/per-item payment schedule. Lady Slipper Designs markets hand-crafted products to retail decorative accessories stores and catalogues. Participants are independent contractors, not employees of the company (Echternach, 1998, personal communication). People choose this type of supplemental income because they can work at home, they can fit the work around existing family responsibilities and the work requires little or no formal education.

Year-round full-time, part-time and summer seasonal jobs are readily available at Anderson Fabrics, a wholesale window treatment, drapery and bedspread business that has a major influence on Beltrami County's economy and its host city, Blackduck. The company began 18 years ago in the basement of the bowling alley with two employees. Currently, Anderson's has a workforce of more than 270, mostly female, a complex of buildings and outlet stores.

Anderson Fabrics is running out of workers, despite a willing pool of single family-head-of-household females. As a recruiting incentive, Anderson Fabrics offers a starting salary of seven dollars an hour, which is about $1.75 above minimum wage. Sheila Lindell, a manager, commented on company policy of paying more than minimum wage, "Can't get them in if you don't pay. They have families."

High-risk Ventures

Some people choose venturesome types of supplementary occupations such as raising exotic animals like ostriches or emu, which requires a risky, more expensive initial cash outlay, yet yields a significant return on their investment. High-risk ventures include other non-traditional diversification besides exotic animals, such as veal operations. The lure of quick profits outweighs the risks. Non-traditional diverse ventures attract people who welcome a challenge, thrive on excitement and have nothing to lose. Risk-takers fall into two categories. The first are retirees who consider high-risk ventures a short-term investment. The second group is young entrepreneurs, who believe that if the bottom falls out of their market, they can always start again.

Wayne Bardwell enjoys danger. Wayne was a crop duster for 20 years. He decided to retire and take up ostrich farming, a physically less dangerous business, but just as risky. Wayne has a flock of about 80 ostriches and is adding to it all the time (1995). The Bardwell family is not unfamiliar with risk. Twenty-five years ago, investing in wild rice paddies was a risky adventure. Warren, Wayne's brother, decided to turn a 405 ha bog into a rice paddy, an unprecedented venture at that time. His foresight and investment paid off. Today, he is one of the largest producers of paddy wild rice in the United States.

In today's economy, even traditional farming can be considered a high risk venture, especially when modified for maximum profit. One farmer, who lives close to the Canadian border, raised veal calves for eight years until prices dropped because of bad publicity.[8] He attributed declining veal consumption to animal activists' protests about cruel living conditions for the calves. He kept his expenses low by purchasing culled bull-calves from local dairy herds. After the Federal Dairy Buy-out of surplus dairy cows in 1988-1989, he switched to a calf-feeder operation and bought culled-calves from Canadian dairies. Calves bought in Canada, even after paying inspections fees and taxes, were less expensive than buying locally. He bought corn from Southern Minnesota and hay from Northern Minnesota and the Dakotas. In 1993, he sold his equipment and livestock at auction. Currently, he drives a truck and does a little logging on the side. His wife is a nurse and writer. She also helped her husband with the farm and, at the same time, commuted 48 km to work each day.

Nostalgia

Nostalgia is the most elusive of the four categories because it is a lifestyle choice rather than a economic strategy, although it emulates traditional rural farming. Rural areas originally developed with an agricultural base, so "even if farming no longer dominates the local economy, vestiges of traits inherent to a community's founding and settling pattern influence its evolution" (Salamon, 1989). Newcomers have the option to adopt rural traits that match their perceptions of an idyllic rural lifestyle. People new to the Boreal Forest attempt Nostalgia farming. Jim and Judy Van deKamp moved to the region five years ago to be nostalgically self-sufficient; yet they cannot achieve this lifestyle without the help of neighbours for hay making, friends to buy their surplus eggs and a daughter who helps with the day-to-day chores. Jim can afford to dabble in nostalgic farming because he receives a pension (Van deKamp, 1992, personal communication).

Innovative Strategies

If an additional classification were added to Northern Minnesota's economic matrix, it would be *Diverse-innovative strategies*. The recycling of tailing ponds and overburdens of the Mesabi Iron Range falls into this category. Local entrepreneurs have converted overburdens into Olympic ski jump training sites.

Mine pits, too, have not escaped recycling. Southern Minnesota experienced a drought in the early 1990s. Several Iron Range townships considered buying abandoned mines and converting mine pits into fresh water reservoirs with the hopes of selling water to drought stricken areas in Southern Minnesota (Bunderman, 1991, personal communication). They scrapped the idea when the cost of environmental clean up outweighed the potential profits. In 1988, Aqua Farms converted mine pits into fresh water trout and salmon farms. The ponds closed in 1993 because of pollution to the water table from excessive fish waste and uneaten food (*The Pioneer*, 1994).

Comments

Contemporary land use in Northern Minnesota is secondary to the residents' occupation. People co-exist with the land, but for most Northern Minnesotans the land does not provide an adequate source of income. Often, people choose land-related occupations because land represents security. The region is evolving from felled forest lands to an amenity-based landscape. The skill of the people and their ability to use marginal resources is the region's primary resource.

The most recent US Census showed that most of the region's population worked less than 40 hours a week in a formal occupation, yet people do work full-time because they combine an odd assortment of part-time jobs to create a year-round self-employment base. While economic choices are limited, there is a great deal of diversity of rural economic activities. Multiple jobs and alternative means of income, rather than being an anomaly, are the norm. Clever ways to make a living result in a mixed economic base. Because Northern Minnesota is a region of extremes, a singular land use and support by a single venture are out of character. What does exist for this region is an alternative economic structure based on ingenuity, but preserved by necessity.

Notes

1. Farming and settlement declined after the 1930s because sub-marginal and marginal land was retired from farming programmes (Committee on Land Utilisation 1934, p. 20), the designation of land as national and state forest (Matson, 1995), a decline of public and private lands with harvestable trees, and changes in land ownership because of tax forfeiture (Committee on Land Utilisation 1934, pp. 131, 163).
2. John Fraser Hart discussed the short recreational season for remote areas in *The Rural Landscape* (1998). He commented that the local people have less than four months in which to milk the tourists of enough to live on for the rest of the year (p. 379).
3. Sources for percentages are from two regional reports, *Overall Economic Development Program Report*, Headwaters Regional Development Commission (1992) and *Overall Economic Development Program Report*, Arrowhead Regional Development Commission (1993). The 1992 US Census of

Agriculture and the 1990 US Census of Population also were used to calculate percentages.

4. *Ibid*. Regional logging numbers were extrapolated from the two economic development reports.

5. Mrs Goslin openly shared her experiences with me because she believed I was born in Northern Minnesota. When she found out that I was originally from California, she said "you're a hippie," and I instantly lost credibility. Amy Goslin's reaction can be partially attributed to her experiences in the 1970s and what she believed to be an invasion of hippies during this time. The US Census shows that the region gained population between 1970 and 1980, but it also shows how fickle people can be with an exodus between 1980 and 1990.

6. A LQ shows an economic enterprise is of greater or lesser importance to the region than to the state or nation as a whole.

7. Lady Slipper Designs began 25 years ago as a Community Action Project in Bemidji, Minnesota, to provide low and moderate-income rural families with an alternative source for economic support. More than 2,000 participants have produced piecemeal handicrafts since 1973. Today, it is one of the leading US cottage industries and has 100 participants, 25 of whom do most of the work (Echternanch, 1998, personal communication).

8. The farmer asked that his name and exact location be withheld.

Chapter 17

Human Capital and Rural Development: What are the Linkages?[1]

Ray D. Bollman

Introduction

Policy analysts, newspaper reporters, best-selling authors and the general public 'believe', in large majority, that the human capacity of the workforce will be the key factor determining improvements in well-being over the medium term. The evidence is scattered and the evidence of interest to rural populations is even more scattered. The objective of this chapter is to assemble, to review and to synthesise the evidence concerning the role of human capacity to improve the well-being of rural people and, by inference, the well-being of rural places.[2]

Why 'Rural'

Rural Canada is experiencing considerable 'demographic pressure' as 1.76 rural persons are now looking for a job for each rural person retiring from the workforce (OECD, 1996, p. 43). Overall, rural areas experience lower employment growth, in part because the fastest growing sector, the business services sector, is largely concentrated in metropolitan centres (Government of Canada, 1995). Rural areas of Canada not adjacent to metropolitan areas are experiencing out-migration, higher unemployment and lower incomes. Thus, there is justification for attention to rural employment policy.

Why 'Human Capital'?

Schultz (1975) has emphasised the value of the ability to deal with disequilibria. As "disequilibria" [read: unpredicted change] confront us from all directions, the ability to "define the problem" and "to solve the problem" takes on a high value. If we merely needed to grow the same crops with the same tools [technology] as our forefathers, then there would be no so-called 'disequilibria' and there would be a low payoff to the human capacity to deal with disequilibria.

'Human capacity' and 'human capital' are used interchangeably in this chapter to encompass the overall capacity of an individual to contribute to his/her own well-being and the well-being of the community/economy. Long treatises have discussed the various components, which include physical health, knowledge, ability to solve problems and even the investment in a geographical move that improves one's earnings. The ability to cope with change and to solve problems is the implicit focus of this paper. The level of formal education is used as a proxy for human capital.

More recently, Reich (1991) has argued that the wealth of a locality is contained in the human capacity of the residents. Wealth in the form of financial assets and technology are easily transferred across borders. The complement of skills of the resident population is the wealth of a locality.

Human Capital: Where Does it Start?

The story starts with evidence of where cognitive skills begin to develop. The importance of nutrition and nurturing of children (starting at age 'minus nine months') has been well documented elsewhere but the explicit link of nutrition and nurturing of children to local economic development is not well documented. Keating and Mustard (1993), Hertzman (1994), Mustard (1994), Nash (1997) and Blakeslee (1997) have reviewed the literature to make a direct linkage between the nutrition and nurturing of children and the ability of a society to generate economic development. Children with good nutrition and good nurturing have the ability to cope and to succeed in a world that now requires an enhanced ability to deal with disequilibria - whether this confronts the child in a Grade One classroom, a high school classroom or when finding or making a job. If there is only one policy investment to be made in the development of human capital, investment in nutrition and nurturing of children should be '*the*' investment.

These arguments are not new. In the 1960s, some analysts (e.g. see Abramson, 1967) suggested that psychological disorders in farm families - due, in part, to isolation and the dashed expectations for good crops and good incomes - fostered learning disabilities and impeded rural development. In such situations, it was not clear if local economic development would be facilitated by public intervention to train the adult population. Rather, attention to nutrition and nurturing the next generation may have been the appropriate target for public intervention. Others (e.g. Popkin, 1972) also documented the relationship between achievement and nutrition.

The Story for Rural Places

Most discussions of future trends predict that 'analytical skills' (i.e. the ability to deal with disequilibria) will provide the big payoff for individuals (and by

208

association, for localities in which these individuals live) (e.g. Reich, 1991). Given the apparent "simultaneous globalisation and localisation" of society (Wade and Pulver, 1991, p. 108), problem-identification and problem-solving skills are needed to participate in the globalising economy. *At the same time, more and more of the responsibility for human capital development is falling to the local level in most jurisdictions.*[3]

Some studies indicate that rural areas with a more highly educated workforce show more development. For example, a study by McGranahan and Kassel (1996) for the OECD has shown that, for a selected group of OECD countries, high-education rural regions showed higher employment growth (or lower employment losses) than low-education rural regions.

Detailed studies in the United States (e.g. McGranahan, 1991; McGranahan and Ghelfi, 1991; Killian and Parker, 1991; Killian and Beaulieu, 1995) show that a simple association between local education levels and local employment growth provides a positive correlation - areas of higher education levels will have higher employment growth. However, simply taking into account the industrial mix and the type of region lowers the impact of an educated workforce to nil. That is, in certain communities with a certain industrial mix, it appears that the presence or absence of an educated workforce has little impact on employment prospects.

These (admittedly) simple models search for the impact of human capital on rural places. The argument is that if a community has a highly-skilled workforce, the jobs will come. The general conclusion from studies in the United States might be summarised anecdotally as being similar to the case of the local community investing in an industrial park - if you do *not* have an industrial park, the jobs will not come; if you do have an industrial park, the jobs still will not come unless you do something more. A well-educated workforce provides a similar benefit (and removes a similar constraint) as an industrial park - it is a necessary but not a sufficient condition.

Thus, localities in the United States that invest in a well-educated workforce should not expect that jobs would come unless they do other things as well. Nevertheless, there remains a high and significant return to individuals to invest in education and training, wherever they end up working.

The Story for Canada

Rural Canada appears disadvantaged. Among OECD countries, Canada has the biggest urban-rural gap in share of the workforce (aged 25 to 44) with university or college graduation (OECD, 1996, p. 170). In census metropolitan areas (CMAs),[4] the share of the population, 15 years of age and over, with less than Grade 9[5] was 12% in 1991 (see Table 17.1). This proportion increases as one moves away from the zone of metropolitan influence and reaches 28% in zones of no metropolitan influence. The share with less than Grade 9 varies considerably among the provinces - in the zones of no metropolitan influence, the share varies from a high of 34% in Québec to a low of 20% in Nova Scotia.

A similar and inverse pattern is shown for the proportion of the population, 15 years of age and over, with Grade 12 or higher years of schooling. The highest shares are in the metropolitan centres (65%) and the lowest shares are in the zones of no metropolitan influence (40%) (see Table 17.2). Again, considerable diversity among the provinces is evident. Thus, the more 'rural' the community, the lower the level of educational attainment in the community.

Table 17.1. Population with less than Grade 9 education, Canada and Provinces, 1991 (percentages)

	CMA/CA	SMIZ	MMIZ	WMIZ	NMIZ	Total
Newfoundland	12	25	27	26	33	20
Prince Edward Island	11	19	19	28	33	15
Nova Scotia	10	15	16	19	20	13
New Brunswick	14	27	26	24	24	20
Québec	18	26	30	28	34	20
Ontario	11	14	16	15	31	12
Manitoba	11	15	23	25	33	15
Saskatchewan	11	17	23	22	24	16
Alberta	7	9	14	14	22	9
British Columbia	8	10	10	11	21	9
Canada	12	17	22	20	28	14

Notes: Key to Zone types: SMIZ - Strong Metropolitan Influence Zone; MMIZ - Moderate Metropolitan Influence Zone; WMIZ - Weak Metropolitan Influence Zone; NMIZ - No Metropolitan Influence Zone.
        A census metropolitan area (CMA) is a core of 100,000 population plus all surrounding municipalities where 50% or more commute into the urban core. A census agglomeration (CA) is a core of 10,000 to 99,999 plus the surrounding municipalities where 50% or more of the workforce commutes into the urban core. In this table, a 'strong' influence zone comprises all municipalities where 20 to 49% of the workforce commutes into a CMA or CA. A 'moderate' influence zone comprises municipalities where 5 to 19% of the workforce commutes into a CMA or CA. A 'weak' influence zone comprises municipalities where >0 to 4.9% of the workforce live in the municipality and work in a CMA or CA.
Source: Statistics Canada, Census of Population, 1991

The federal government, as part of its 1991 'Prosperity Initiative', proposed an objective of having 90% of all individuals having Grade 12 (or equivalent) by age 25 (Statistics Canada, 1991). In 1991, only 11% of census consolidated sub-divisions[6] reported that over 90% of residents aged 20 to 24[7] had achieved Grade 12 or equivalent schooling (see Table 17.3). As indicated above, the incidence with less than Grade 12 is higher in rural regions of Canada.

One reason for a lower attainment of higher education in rural regions is a (perceived and perhaps real) lower demand for workers with higher education in rural areas. There is consequently a (perceived and perhaps real) lower pay-off to higher education in rural regions. Looker (1997) found that rural young people aspire to *and* attain a lower level of education. Interestingly, many rural young people do aspire to courses that 'train' entrepreneurs. Hajesz and Dawe (1997) found that over two-thirds of this population group in their sample would take an entrepreneurship class if taught in the school and half would take an entrepreneurship class if taught outside the school.

Table 17.2 Population with Grade 12 or higher education, Canada and Provinces, 1991 (percentages)

	CMA/CA	SMIZ	MMIZ	WMIZ	NMIZ	Total
Newfoundland	62	43	41	45	34	51
Prince Edward Island	65	52	49	39	19	57
Nova Scotia	62	51	52	49	43	57
New Brunswick	62	46	46	49	48	55
Québec	64	52	47	48	40	61
Ontario	65	57	53	55	38	64
Manitoba	61	51	42	42	32	55
Saskatchewan	62	48	44	45	41	54
Alberta	68	57	52	53	45	64
British Columbia	67	60	59	58	46	66
Canada	65	54	49	50	40	62

Notes: For explanation of zone types see Table 17.1
For explanation of categorisation of Census Areas see Table 17.1
Source: Statistics Canada, Census of Population, 1991

Table 17.3. Completion of high school education for youths 20 to 24 years, Canada, 1991 (counts)

Percentage of youths 20 to 24 years who have completed high school	Number of Census Consolidated sub-divisions	Percentage
Less than 50	335	14.0
50 to 60	193	8.0
60 to 70	449	19.0
70 to 80	667	28.0
80 to 90	501	21.0
90 to 95	89	4.0
Over 95	171	7.0
Total	2,405	100.0

Notes: Completed high school includes equivalent (i.e. have taken post-secondary training).
Analysis based on total number of Census Consolidated Sub-divisions with 40 or more individuals aged 20 to 24 years.
Source: Statistics Canada, Census of Population, 1991

The Impact of the Level of Schooling on Local Economic Development

A Preliminary Model

As noted above, the literature for the United States provides little support for the hypothesis that higher levels of schooling will bring about local job growth. However, this relationship has received little attention in Canada. The purpose of this section is to develop a preliminary model of local economic development to evaluate the contribution of schooling levels to local development.

The research attempting to explain international, national and local economic development is vast and varied. Issues of employment demand including technological change and issues of employment supply including labour mobility must be considered. Research attempting to explain differences in growth among countries generally finds that the level of human capital in the initial period positively influences subsequent national growth but the growth in the level of human capital appears to contribute little to national economic growth (Griliches, 1996). Research to explain employment growth among US counties appears to indicate that community education levels have no impact if one simply controls for the mix of employment by industrial sector as an explanatory factor in local employment growth in the US (e.g. Killian and Parker, 1991, p. 108).

Freshwater *et al.* (1996) developed a simultaneous three-equation model to estimate the impacts of different variables on development outcomes and to test whether areas within the Tennessee Valley Authority showed higher levels of development outcomes. In effect, Freshwater *et al.* (1996) acknowledge that development is not a univariate phenomenon. They propose three measures of development, they develop an equation to explain each measure, and they explicitly recognise the endogeneity among the three measures by including each of the other two measures in their three equations (estimated simultaneously by 3-stage least squares). A similar analysis is planned for Canada in the future, but the results reported here start with a single-equation ordinary least squares model in the spirit of Kusmin *et al.* (1996).

Following Freshwater *et al.* (1996), development is recognised as multi-dimensional. Development policy pursues more than one objective. Community welfare is measured in more than one dimension. We offer four measures of local community 'development'[8] that are admittedly narrowly focused on the performance of the labour market:

1. the rate of growth[9] of average real[10] earnings[11] per worker in the community (for individuals with earned income, 15 years of age and over) (LNCAVERN)[12];
2. the rate of growth of average real hourly wage rates[13] for workers in the community (LNCWAGE);
3. the rate of growth of employment in the community (LNCEMP); and
4. the rate of growth of community aggregate earnings[14] (LNCTEARN) is offered as a comprehensive indicator of community economic development.

Kusmin *et al.* (1996) argue that the growth in community aggregate earnings (whether due to employment growth, or growth in earnings per worker, or both) is a useful single indicator of local economic development.

In this chapter four sets of factors are identified to explain growth within localities:

1. a measure of the level of **human capital** in the community;
2. variables capturing the mix of **employment by industrial sector**, as local economic development will be (dis)advantaged by whether the local industrial sector is concentrated in expanding (declining) sectors;
3. measures of **local factors** influencing the level of local development; and
4. variables to capture the **nature of the region** within which the local economy is situated.

The Level of Local Human Capital

Two alternative ways of measuring the level of local human capital are tested:

YOS1981: the average years of schooling for all individuals 15 years of age and older in the community; and
LTGR981: the percentage of individuals, 15 to 64 years of age, with less than Grade 9 in 1981; plus
SOMEU81: the percentage of individuals, 15 to 64 years of age, with some post-secondary schooling in 1981.

The idea here is that localities with low levels of education may gain due to increases in low-tech manufacturing jobs and, at the same time, localities with well educated workforces may gain employment as they were more capable of participating in 'new economy' jobs.

Mix of Employment by Industrial Sector

The share of employment in four key sectors with (potentially) exportable goods and services is included:

PRIM81: the percentage of employment in 1981 in agriculture, fishing, forestry, mining and oil extraction, and hunting and trapping.
TRMFG81: the percentage of employment in 1981 in traditional manufacturing activities (manufacturing industries not designated below as 'complex').
COMFG81: the percentage of employment in 1981 in complex manufacturing activities (includes printing/publishing, machinery, aircraft, electrical products, petroleum and coal products, chemicals and scientific/professional equipment industries).
PRSERV81: the percentage of employment in 1981 in producer service activities (i.e. finance, insurance, real estate and business services such as accounting, consulting, software design and development, etc.).

Only the shares of employment in sectors with (potentially) exportable goods and services are identified. These employment sectors are generally driven by market demand from outside the community. Most other sectors are driven by market demand from within the community and thus are endogenous with local population growth. Recognition of the dependency of the community on key types of export markets is expected to explain part of local economic development in the 1980s. Specifically, the share of employment in primary sectors and in traditional manufacturing is expected to be negatively associated with both growth in employment and growth in earnings per worker. Specialisation in complex manufacturing and producer services is expected to be positively associated with employment growth and earnings growth.

Local Factors Influencing Local Economic Development

Eight measures of quantifying the influence of local factors on economic development are tested:

SELF81: the percentage of individuals, 15 to 64 years of age, who were self-employed in 1981 (excluding farm self-employed). One hypothesis would suggest that areas with a higher share of self-employed in the labour force (i.e. more 'entrepreneurs') would generate more employment growth (but growth in average earnings would be expected to lag). An alternate hypothesis would suggest that a high incidence of self-employment in the initial period is an indicator that there is little prospect for new wage jobs and thus unemployed workers have resorted to self-employment endeavours.

ABORIG81: the percentage of individuals in 1981 with an Aboriginal ethnic background. Localities with a higher share of Aboriginal population show a boom in the Aboriginal working age population due to the high fertility rates in the last two decades. However, an expanding potential work force may not translate into expanding employment. Typically, unemployment is higher, labour force participation rates are lower and outward mobility is significant. Growth in average earnings would be expected to be less than average.

UNEMP81: the percentage of the labour force, 15 to 64 years of age, which is unemployed in 1981. A high unemployment rate indicates an excess supply of labour that would be expected to generate employment growth but growth in earnings per worker would be expected to be lower.

EDUCIN81: the percentage of individuals in 1981, 15 to 64 years of age, working in the 'educational industry' (i.e. working in an educational institution), either as an instructor or as support staff. It is expected that the knowledge infrastructure provided by the members of the educational industry would provide a positive contribution to local employment growth. However, the education industry *per se* achieved significant earnings growth in the 1970s even **relative** to the significant real growth in all sectors. Thus, it is expected that localities with a higher share of employment in education industries would show lower earnings growth in the 1980s.

YOUTHIN81: the percentage of youth in 1981, aged 25 to 29, who have moved into the locality in the five years previous to 1981. This might be interpreted as an indicator of 'expected' growth over the subsequent period. It is also expected to augment the level of human capital. Both employment growth and average earnings growth are expected to be positively associated with this variable.

PEROLD81: the percentage of the population in 1981 that is 55 to 74 years of age. A high share of individuals in this age class usually results from the outward mobility of youth and thus this variable is an indicator of 'expected' employment decline and/or earnings decline. Also, a high share of the population in this age category is expected to have lower educational levels and thus this variable is also intended to account for the fact that a measured low average educational attainment level may result from a high share of older persons in the community.

NEW5581: the percentage of individuals 55 to 74 year of age in the locality who moved into the locality in the 5 years prior to 1981. This is a proxy for a retirement destination community. When this factor is strong, employment growth is expected to provide services for the retirees. The impact on earnings per worker is uncertain as the new jobs may be in lower-paying service sector jobs.

LT21K81: the percentage of individuals in the locality living in households with income less than the national median of $21,000 (current 1981 dollars). There is a new and expanding literature (see Osberg, 1995) that suggests that places with a more equal distribution of welfare will experience more growth. In other words, there is not a trade-off between equity and growth - rather a more equal distribution of welfare contributes to economic growth. It is expected that a larger share of low-income individuals will reduce local employment growth. However, localities with a high incidence of low incomes in 1981 would be expected to regress towards the mean and thus to show relatively higher earnings growth in the 1981 to 1991 period.

The Nature of the Region in which the Locality is Situated

The typology developed by Hawkins (1995) (see also Hawkins and Bollman, 1994; Bollman, 1994) is used to indicate the nature of the region. Seven types of regions were identified. Dummy variables are used to indicate the type of region in which the locality is situated. Census divisions with large cities were identified as primary settlements (DPRSETTL). Census divisions with smaller cities were identified as urban frontier (DURBFRON). The excluded class[15] of census divisions were the rural nirvana census divisions - these were rural census divisions benefiting from the metropolitan influence of Toronto plus a few census divisions around Vancouver, Winnipeg and Montreal. Census divisions where agriculture was important were agro-rural (DAGRRUR). These were located in the grain belt of Saskatchewan and Manitoba plus the agricultural areas of Québec. Census divisions with poor economic prospects were clustered as rural enclave census divisions (DRURENCL). These census divisions include Pontiac County in western Québec plus most census divisions in the Gaspé region of Québec, northern New Brunswick, the ends of Prince Edward Island, the ends of Nova Scotia and outport Newfoundland. Census divisions endowed with natural resources (forestry, mining, oil and gas) including many Alberta census divisions plus northern British Columbia and northern Ontario and those with good human capital resources in the capital cities of Whitehorse, Yukon and Yellowknife, Northwest Territories were labelled resourced areas (DRESAREA). Other northern census divisions with a larger Aboriginal population were designated as native north census divisions (DNATIVNO).

The excluded class was labelled 'rural nirvana' because these regions were experiencing both population growth and earnings growth as they were in the rural shadow of large metropolitan markets. Thus, DPRSETTL may be expected to have higher employment growth and higher earnings growth than the excluded regions and all other regions would be expected to have lower employment and lower earnings growth.

The association between these variables and the four measures of community development have been estimated for three sets of observations:

1. all communities in all regions (i.e. all census consolidated sub-divisions in all census divisions in Canada);
2. all communities in only *predominantly rural regions*, as defined by the OECD (1996) (i.e. census divisions with more than 50% of their population living in rural communities);
3. only rural communities, as defined by the OECD (1996), regardless of the type of region in which they are situated (i.e. all census consolidated sub-divisions with less than 150 inhabitants per square kilometre).

The three sets of results are discussed to see if the relationships also apply when the analysis is restricted to communities in rural regions and to rural communities, regardless of the type of region.

Results of Census Data Analysis

The results for four regression equations have been estimated by ordinary least squares using data tabulated from the 1981 and 1991 Censuses of Population. The objective is to determine the empirical association between community human capital and the growth in real average earnings (LNCAVERN), the growth in real average hourly wages (LNCWAGE), the growth in community employment (LNCEMP), and the growth in aggregate community earnings (LNCTEARN).[16] The equations where community human capital is measured by **average** years of schooling (YOS1981) are presented in Table 17.4.

The equations where the community human capital complement is indicated by the **distribution** of the population by level of education attainment are presented in Table 17.5. The adjusted R^2 ranges from 0.21 to 0.35, which is consistent with similar studies investigating local community growth. There is no difference in adjusted R^2 between equations with the average years of schooling (YOS1981 in Table 17.4) and the distribution of educational attainment (LTGR981 and SOMEU81 in Table 17.5).

216

Table 17.4. Factors associated with local economic development, Canada, 1981 to 1991

Dependent variable:	Mean	LNCTEARN mean=10.63 Reg Coeff	t-stat	LNCAVERN mean=-3.77 Reg Coeff	t-stat	LNCWAGE mean=-0.03 Reg Coeff	t-stat	LNCEMP mean=12.29 Reg Coeff	t-stat
Constant		59	4.5	-5.2	-0.6	0.2	1.8	22	2.2
Level of local human capital									
YOS1981	10.52	-1.4	-1.3	1.40	1.9	-0.01	-1.0	0.86	1.1
Industrial structure of employment by industry									
PRIM81	20.72	-0.68	-13.0	-0.36	-9.4	-0.00	-7.7	-0.3	-7.0
TRMFG81	14.24	-0.18	-2.9	-0.01	-0.2	0.00	1.9	-0.31	-5.4
COMFG81	2.18	0.24	1.1	-0.01	0.1	0.00	1.4	0.2	1.1
PRSERV81	3.93	0.21	1.0	-0.14	-1.1	-0.00	-0.3	0.28	1.5
Local factors influencing the dependent variable									
SELF81	5.22	-0.04	-0.2	-0.09	-0.5	0.00	1.5	0.06	0.3
ABORIG81	2.24	-0.32	-1.9	0.04	0.8	-0.00	-0.9	-0.24	-1.7
UNEMP81	9.34	-0.40	-4.5	-0.05	-0.9	-0.00	-2.5	0.016	0.2
EDUCIN81	5.55	-0.58	-3.1	-0.56	-4.8	-0.01	-4.3	-0.16	-1.0
YOUTHIN81	36.95	-0.01	-0.4	-0.03	-1.1	-0.00	-1.4	-0.00	-0.1
PEROLD81	15.96	-0.93	-6.8	-0.15	-1.6	-0.00	-3.3	-0.7	-6.5
NEW5581	10.73	0.41	5.7	-0.02	-0.3	0.00	0.5	0.44	6.9
LT21K81	51.67	0.34	7.3	0.24	6.8	0.00	5.8	0.08	2.1
Type of region in which the locality is located									
DPRSETTL	0.01	-8.9	-1.8	-6.8	-3.4	-0.05	-2.4	-5.7	-1.4
DURBFRON	0.28	-19	-11.0	-11	-10.6	-0.10	-8.3	-7.7	-5.7
DAGRRUR	0.32	-25	-14.0	-14	-13.8	-0.17	-13.6	-6.5	-4.7
DRURENCL	0.12	-18	-7.6	-9.9	-6.5	-0.04	-2.2	-14	-7.2
DRESAREA	0.07	-15	-5.5	-7.2	-4.7	-0.04	-2.3	-8.7	-3.8
DNATIVNO	0.01	-7	-0.8	-12	-2.6	-0.05	-1.4	-1.1	-0.1
No. of obns		2315		2315		2315		2315	
Adjusted R^2		0.34		0.22		0.22		0.21	

Notes: All t-statistics have been corrected for heteroscedasticity
See text for explanation of the variable names

Table 17.5. Factors associated with local economic development, Canada, 1981-1991

Dependent variable	Mean	LNCTEARN mean=10.63 Reg Coeff	t-stat	LNCAVERN mean=-3.77 Reg Coeff	t-stat	LNCWAGE mean=-0.03 Reg Coeff	t-stat	LNCEMP mean=12.29 Reg Coeff	t-stat
Constant		36	5.5	12	2.7	0.10	1.8	25	4.5
Level of local human capital									
LTGR981	24.37	0.31	2.8	-0.05	-0.5	0.00	1.1	0.13	1.5
SOMEU81	27.02	0.07	0.6	-0.06	-0.7	-0.00	-0.5	0.17	1.8
Industrial structure of employment by industry									
PRIM81	20.72	-0.68	-12.8	-0.36	-9.4	-0.00	-7.7	-0.29	-6.7
TRMFG81	14.24	-0.21	-3.4	-0.03	-0.7	0.00	1.6	-0.32	-5.7
COMFG81	2.18	0.27	1.2	0.05	0.5	0.00	1.6	0.21	1.2
PRSERV81	3.93	0.22	1.1	-0.04	-0.3	-0.00	-0.1	0.28	1.5
Local factors influencing the dependent variable									
SELF81	5.22	-0.05	-0.2	-0.09	-0.5	0.00	1.4	0.05	0.2
ABORIG81	2.24	-0.32	-1.9	0.03	0.6	-0.00	-0.9	-0.25	-1.8
UNEMP81	9.34	-0.43	-5.0	-0.09	-1.5	-0.00	-2.8	-0.01	-0.1
EDUCIN81	5.55	-0.61	-3.2	-0.51	-4.2	-0.01	-4.2	-0.18	-1.1
YOUTHIN81	36.95	-0.01	-0.2	-0.02	-0.7	-0.00	-1.2	-0.00	-0.1
PEROLD81	15.96	-0.93	-6.8	-0.14	-1.4	-0.00	-3.3	-0.70	-6.5
NEW5581	10.73	0.43	5.9	-0.01	-0.2	0.00	0.6	0.45	7.1
LT21K81	51.67	0.32	7.0	0.22	6.3	0.00	5.7	0.07	1.9
Type of region in which the locality is located									
DPRSETTL	0.01	-9.6	-1.9	-6.1	-3.0	-0.05	-2.4	-6.4	-1.5
DURBFRON	0.28	-19	-11.0	-11	-10.1	-0.10	-8.0	-8.5	-6.0
DAGRRUR	0.32	-26	-14.3	-15	-13.8	-0.17	-13.5	-7.4	-5.3
DRURENCL	0.12	-18	-7.8	-10	-6.4	-0.04	-2.1	-16	-7.7
DRESAREA	0.07	-16	-5.8	-7.6	-4.9	-0.04	-2.4	-9.4	-4.1
DNATIVNO	0.01	-9.4	-1.0	-13	-2.8	-0.06	-1.4	-4	-0.5
No. of obns		2315		2315		2315		2315	
Adjusted R^2		0.35		0.22		0.22		0.21	

Notes: All t-statistics have been corrected for heteroscedasticity
See text for definition of the variable names

The results indicate that the association between the measures of human capital and of community economic development are generally weak.[17] There was a weak association between the community average years of schooling (YOS1981) and both a **higher** rate of growth of community employment (LNCEMP) and a **higher** rate of growth of average earnings per worker (LNCAVERN) (as summarised in Table 17.6). On the other hand, there was a weak association between the community average years of schooling (YOS1981) and a **lower** rate of growth of community average hourly wage rates (LNCWAGE). Overall, as a result, there was a weak association between the community average years of schooling (YOS1981) and a **lower** rate of growth of aggregate community earnings (LNCTEARN). Evidently, community aggregate earnings (LNCTEARN) grew less in communities with a higher level of average education (YOS1981) because the lower growth in wages (LNCWAGE) was not off-set by the growth of employment (LNEMP) and/or the growth of the hours worked component of the growth in average worker earnings (LNCAVERN).

Our alternative measures of the community's human capital complement consider the distribution of the population by level of educational attainment. The results indicate that areas with a lower educational attainment (LTGR981) and areas with higher education attainment (SOMEU81) were both weakly associated with a **higher** rate of growth of employment (LNCEMP) in the 1980s. Interestingly, communities with a higher share of **lower** educated individuals (LTGR981) were weakly associated with **higher** growth of wages (LNCWAGE), but when combined with the weak association with a **higher** rate of employment growth, we find communities with a higher share of their population with a lower education (LTGR981) have a significant association with **higher** growth in aggregate community earnings (LNCTEARN). Thus, it appears that communities with low-skilled workers (as indicated by a high share of individuals with a lower level of education) were able to attract jobs during the 1980s and were also able to increase their wage level during this period.

Many of the other variables have the hypothesised association with the dependent variables. As expected, employment specialisation in the primary sectors (PRIM81) was associated with **lower** growth in all measures of community development outcomes. Community specialisation in traditional manufacturing (TRMFG81) was significantly associated with **lower** employment growth (LNCEMP) and with **lower** growth in aggregate community earnings (LNSTEARN). Community employment specialisation in the primary and traditional manufacturing sectors constrained the growth in community development outcomes during the 1980s.

The share of the labour force that was self-employed (SELF81) had a weak association with a higher growth of hourly wages (LNCWAGE), contrary to expectations. However, there was no significant association with job growth. Thus, communities with a higher share of the workforce being self-employed did not indicate an 'entrepreneurial' community and with higher job growth.

A higher share of Aboriginal people in the population (ABORIG81) was weakly associated with **lower** employment growth (LNCEMP) and weakly associated with **lower** growth in community aggregate earnings (LNCTEARN). In

Table 17.6. Comparison of hypothesised and actual results

	Growth in (constant $) aggregate community earnings (LNCTEARN)		Growth in (constant $) average earnings per worker (LNCAVERN)		Growth in (constant $) hourly wage rate (LNCWAGE)		Growth in Community employment level (LNCEMP)	
	Exp'd	Actual	Exp'd	Actual	Exp'd	Actual	Exp'd	Actual
Level of local human capital								
YOS1981	+	neg.	+	pos.	+	neg.	+	pos.
LTGR981	?	**POS.**	-	n.s.	-	pos.	+	pos.
SOMEU81	+	n.s.	+	n.s.	+	n.s.	+	pos.
Industrial structure of employment by industry								
PRIM81	-	**NEG.**	-	**NEG.**	-	**NEG.**	-	**NEG.**
TRMFG81	-	**NEG.**	-	n.s.	-	pos.	-	**NEG.**
COMFG81	+	pos.	+	n.s.	+	pos.	+	pos.
PRSERV81	+	pos.	+	n.s.	+	n.s.	+	pos.
Local factors influencing the dependent variable								
SELF81	?	n.s.	-	n.s.	-	pos.	+	n.s.
ABORIG81	?	neg.	-	n.s.	-	n.s.	+	neg.
UNEMP81	?	**NEG.**	-	n.s.	-	**NEG.**	+	n.s.
EDUCIN81	?	**NEG.**	-	**NEG.**	-	**NEG.**	+	neg.
YOUTHIN81	+	n.s.	+	n.s.	+	neg.	+	n.s.
PEROLD81	-	**NEG.**	-	neg.	-	**NEG.**	-	**NEG.**
NEW5581	?	**POS.**	?	n.s.	?	n.s.	+	**POS.**
LT21K81	?	**POS.**	+	**POS.**	+	**POS.**	-	pos.
Type of region in which the locality is located								
DPRSETTL	?	neg.	?	**NEG.**	?	**NEG.**	?	neg.
DURBFRON	?	**NEG.**	?	**NEG.**	?	**NEG.**	?	**NEG.**
DAGRRUR	-	**NEG.**	-	**NEG.**	-	**NEG.**	-	**NEG.**
DRURENCL	-	**NEG.**	-	**NEG.**	-	**NEG.**	-	**NEG.**
DRESAREA	?	**NEG.**	?	**NEG.**	?	**NEG.**	?	**NEG.**
DNATIVNO	-	n.s.	-	**NEG.**	-	neg.	-	n.s.

Notes: n.s. indicates not significant (with a t-statistic less than 1.0);
pos. or neg. indicates the sign of the association and a t-statistic between 1.0 and 2.0;
POS. and **NEG**. indicates the sign of the association and a t-statistic of 2.0 or more; see text for explanation of the variable names

spite of a high demographic demand for jobs in Aboriginal communities, job growth was less in the 1980s, holding all other factors constant.

A higher rate of unemployment (UNEMP81) in the community in the initial period constrained the growth of hourly wage rates (LNCWAGE), as expected. The apparent excess supply of labour in the initial period, as indicated by a higher rate of unemployment, had no significant impact on job growth in the subsequent period. Communities with a higher share of employment in the education sector (EDUCIN81) experienced lower employment growth (LNCEMP), lower wage growth (LNCWAGE), a lower growth in average earnings (LNCAVERN), and consequently, a lower growth in aggregate community earnings (LNCTEARN) in the 1980s. The presence of an educational institution did not spur local economic development.

A higher share of in-migration by youth (YOUTHIN81) in the previous period was not associated with employment growth or with earnings growth. A higher share of in-migration by youth was weakly associated with **lower** wage growth (LNCWAGE). It was hypothesised that this variable would signal areas expected to grow in the subsequent period. A higher share of older individuals in the population (PEROLD81) did signal past out-migration of youth and was associated with **lower** employment growth (LNCEMP) and with **lower** earnings growth (LNCAVERN and LNCTEARN). Retirement-destination communities (NEW5581) appear to generate significant growth in aggregate community earnings (LNCTEARN) by generating significant growth in employment (LNCEMP). Communities with a higher share of poor persons (LT21K81) were associated with **higher** growth in all measures of community economic development. It appears that these communities were catching up (i.e. had relatively higher growth rates) during the 1980s.

The type of region in which the community is located does matter. Communities in each type of region showed less growth than communities in the omitted category - the booming 'rural nirvana' regions. Note that employment growth was much lower in communities in the 'rural enclave' (DRURENCL) regions, relative to the communities in the excluded group. Regarding growth in wages (LNCWAGE), growth in average earnings (LNCAVERN) and growth in community aggregate earnings (LNCTEARN), we see the lowest growth (i.e. the largest negative coefficient) for communities in the 'agro-rural' regions (DAGRRUR).

Recall that the United States studies found no significant association between community employment growth and community education levels, if the industrial structure of employment and the type of region were taken into account. In this study, we have controlled for the industrial structure of employment and the type of region and we do obtain a positive (albeit weak) association between employment growth and education levels.

The results above refer to all communities in all regions. The equations presented in Table 17.4 and Table 17.5 were also estimated for communities in *predominantly rural* regions and for all rural communities, regardless of the type of community. We discuss the results regarding the human capital variables here. In general, the association between our measures of community human capacity (YOS1981, LTGR981, SOMEU81) and community economic development

221

outcomes are consistent with the discussion above (see Table 17.7). This contrasts with the results of Killian and Parker (1991) who found no association between employment growth and education levels for rural areas. However, the association here is not strong. The weak association between average levels of schooling (YOS1981) and **higher** employment growth (LNCEMP) holds for communities in rural regions and for rural communities, regardless of the type of region. Interestingly, the association between employment growth (LNCEMP) and the distribution of the population by educational attainment (LTGR981 and SOME81) also holds when we constrain our analysis to communities in *predominantly rural* regions and to rural communities, regardless of the type of region. Specifically, for both rural communities and communities in rural regions, it is evident that communities with a higher share of their population with a lower educational attainment **and** communities with a higher share of their population with a higher educational attainment are both associated, weakly, with **higher** employment growth. Some rural communities with lower skilled workers were able to attract jobs, as were some with higher skilled workers.

Conclusions

The equations explained only 21 to 34% of the variability in local community development in the 1980s. Contrary to the research in the United States, these findings suggest that the human capital complement in Canada's rural communities did provide a positive (albeit weak) boost to job growth in the locality during the 1980s. However, after the lower wage growth is taken into account, aggregate community earnings grew less in communities with a higher level of education. Thus, what are the linkages between human capital and rural development? First, the literature suggests human capacity is largely developed by the nutrition and nurturing of children, specifically in the period of minus nine months to plus three years. There is mounting physiological evidence of the linkage between the nutrition and nurturing of infants and their subsequent ability to cope and adapt. Arguably, this is the first place that localities should focus their attention on human capacity development. Secondly, a higher human capacity in a community (as proxied by years of schooling) is weakly associated with a higher growth in community employment but is weakly associated with a lower growth in wages that appears to cause a weak association with lower aggregate community earnings. Investment in nutrition and nurturing of children is a key factor. A higher education level in a community provided only a weak employment boost during the 1980s.

Improving the human capacity of the local workforce is essential to provide opportunities for the individuals, regardless of where they will work. Although human capital resources are essential to participate in the new globalizing economy, local economic development strategies should recognise they need to focus on more than human capital development to stimulate local economic development. As noted by von Meyer and Muheim (1997) and von Meyer (1997):

"The success of the dynamic rural regions is not due to favourable sectoral mixes. The positive performance in creating rural employment results from specific territorial dynamics that are not yet properly understood, but probably include aspects such as regional identity and entrepreneurial climate, public and private networks, or the attractiveness of the cultural and natural environment." (von Meyer, 1997, p. 20)

Table 17.7. Summary of association between community education and community development outcomes, Canada, 1981 to 1991

	Growth in (constant $) Aggregate Community Earnings	Growth in (constant $) Average Earnings per Worker	Growth in (constant $) Hourly Wage Rate	Growth in Community Empl't Level
No. of Communities	LNCTEARN	LNCAVERN	LNCWAGE	LNCEMP

ALL REGIONS:
Average years of schooling in the beginning period (YOS1981)

2,315	(negative)	(positive)	(negative)	(positive)

Share with less than Grade 9 education in the beginning period (LTGR981)

2,315	POSITIVE	n.s.	(positive)	(positive)

Share with some university education in the beginning period (SOMEU81)

2,315	n.s.	n.s.	n.s.	(positive)

PREDOMINANTLY RURAL REGIONS:
Average years of schooling in the beginning period (YOS1981)

1,784	NEGATIVE	n.s.	(negative)	(positive)

Share with less than Grade 9 education in the beginning period (LTGR981)

1,784	POSITIVE	n.s.	(positive)	(positive)

Share with some university education in the beginning period (SOMEU81)

1,784	(positive)	n.s.	n.s.	POSITIVE

RURAL COMMUNITIES:
Average years of schooling in the beginning period (YOS1981)

2,170	(negative)	(positive)	(negative)	(positive)

Share with less than Grade 9 education in the beginning period (LTGR981)

2,170	POSITIVE	n.s.	(positive)	(positive)

Share with some university education in the beginning period (SOMEU81)

2,170	n.s.	n.s.	n.s.	(positive)

Note: n.s. indicates not significant (with a t-statistic less than 1.0);
pos. or neg. indicates the sign of the association and a t-statistic between 1.0 and 2.0;
POSITIVE and NEGATIVE indicates the sign of the association and a t-statistic of 2.0 or more;
See text for explanation of the variable names

Notes

1. This research was initiated when the author was the Stanley Knowles Visiting Professor at Brandon University. Earlier versions were presented to the Second National Congress on Rural Education, Saskatoon, Saskatchewan, 19-21 February, to the 48th European Association of Agricultural Economists Seminar on Rural Restructuring within Developed Economies, Dijon, France, 20-21 March 1997, and to the Joint Meetings of the Canadian Agricultural Economics and Farm Management Society and the American Agricultural Economics Association, 28-30 July 1997, Toronto.

2. The not uncommon observation that a community may be dying but the few remaining individuals may report above average levels of well-being is acknowledged.

3. However, some countries that were not able to provide detailed data did report better employment growth (or less loss) in low-education rural regions.

4. A census metropolitan area (CMA) is a city with an urban-core population of 100,000 or more plus the population in all surrounding municipalities where more than 50 percent of the workforce commutes into the urban core.

5. The 'grade' level indicates the number of years of formal education. The almost universal progression is one grade level per year and children in Grade 1 are typically 6 years of age. In terms of the ISCED (International Standard Classification of Education), Grade 9 is equivalent to a lower secondary education (ISCED level 2) and Grade 12 is equivalent to an upper secondary education (ISCED level 3).

6. A census consolidated sub-division (CCS) is an incorporated municipality, township, town or city. If a small incorporated town is surrounded by a municipality, the two are 'consolidated' for statistical purposes as a CCS.

7. The data for residents aged 20 to 24 are reported to indicate the performance of the educational system for the age group that was the most recent potential group of high school graduates **and** to assess the status of CCS educational attainment relative to the government objective of a 90% high school graduation rate by age 25.

8. Freshwater *et al.* (1996) explicitly recognise that the level of human capital in a locality is (may be) a desired developmental outcome in its own right (this was their third measure of development) **and at the same time**, they want to know the role of this human capital in promoting the levels of the other indicators of development.

9. In each case, the rate of growth is calculated as the difference of the logarithm of the levels: ln (1990 level) minus ln (1980 level).

10. The rate of growth of *real* earnings and *real* wage rates are observed by first deflating the 1990 data to 1980 data before calculating the rate of growth.

11. 'Earnings' includes wages and salaries plus net self-employment income from operating a farm or non-farm business. Conceptually, this is equivalent to multiplying the hourly wage rate times the number of hours worked.

12. Each variable is identified in bold by an acronym (in this case, the acronym represents the logarithm of the change in average earnings) that is used in the tables and the subsequent discussion.

13. The hourly wage rate is estimated as the wages and salaries plus net self-employment income reported for the previous year divided by (the number of hours worked in the week prior to the census multiplied by the number of weeks worked in the previous year).

14. Community aggregate earnings is calculated as the sum of 'earnings' for each individual who resides in the community.

15. Dummy variables (i.e. 0,1 variables) are used to indicate into which group an observation is classified. One group must be excluded from the analysis to prevent the matrix from being singular (which is caused by the sum of the dummy variables for each observation being equal to 1). The coefficient on each dummy variable indicates the impact on the dependent variable by the given dummy variable, **relative to the excluded group**.

16. These equations essentially assume that the local community is in a disequilibrium situation in the initial period. The level of each determining variable in the initial (1981) period is hypothesised to influence the change in each measure of community development toward an equilibrium state in the subsequent period. See Newman and Sullivan (1980) for a detailed discussion of alternative frameworks.

17. The nature of the association between education and community development outcomes differs somewhat due to the exact specification of the equations. The results presented here **do not overstate** the role of human capital in community economic development.

References

Abler, R., Adams, J.S. and Gould, P. (1971) *Spatial Organization: the Geographer's View of the World.* Englewood Cliffs: Prentice-Hall.

Abramson, J.A. (1967) *Rural Non-farm Communities and Families: Social Structure, Process and Systems in Ten Saskatchewan Villages.* Saskatoon: The Canadian Centre for Community Studies, University of Saskatchewan.

Algeo, K. (1997) The rise of tobacco as a Southern Appalachian staple: Madison County, North Carolina, *Southeastern Geographer* 37, pp. 46-60.

Alinski, S. (1972) *Rules for Radicals.* New York: Vintage Books, Random House.

Allsop, D. (1993) Pub guides with pulling power, *The Sunday Telegraph Review* 16 May, p. 15.

The Alzheimer's Disease Society (1992) *The Alzheimer's Disease Report.* London.

Anding, T.L., Adams, J., Casey, W., de Montille, S. and Goldfein, M. (1990) *Trade Centers of the Upper Midwest: Changes from 1960 to 1989.* Minneapolis, Minnesota: Center for Urban and Regional Affairs, Hubert H. Humphrey Center.

Andrews, F.M. and Withey, S.B. (1976) *Social Indicators of Well-Being.* New York: Plenum Press.

Annis, R.C., Everitt, J.C., Bessant, K. and McGuinness, F. (1994) *Sustainable Prairie Communities: Developing a New Agenda.* Brandon: Brandon University, WESTARC Group Inc.

Appalachian Land Ownership Task Force (1983) *Who Owns Appalachia?* Lexington: The University Press of Kentucky.

Appalachian Regional Commission (1984) Displaced workers: Appalachian case studies, *Appalachia* 17 (5-6), pp. 35-37.

Appalachian Regional Commission (1985) Appalachia: twenty years of progress, *Appalachia* 18 (3), pp. 1-108.

Appalachian Regional Commission (1998) *Economic Status of Appalachia and the Mississippi Delta* (Map of regional commission areas). Washington: Appalachian Regional Commission.

Aquino, J. (1993) The politics of landfills, *Waste Age* March, pp. 37-40.

Arkleton Trust (1989) *Appraisal of the Factors which influence the Evolution of Agricultural Structures in the Community and contribute to the Efficiency of the Common Agricultural Policy at the Regional and Farm Level.* Enstone, Oxfordshire: Arkleton Trust.

Arrowhead Regional Development Commission (1993) *Overall Economic Development Program Report.*

Ashton, W. (1990) *Integrative Planning Praxis.* Sackville, New Brunswick: Rural and Small Town Programme, Mount Allison University.

Ashton, W. and Lightbody, J. (1990) Reforming Alberta's municipalities: possibilities and parameters, *Canadian Public Administration* 33 (4), pp. 506-525.

Ayres, J., Cole, R., Hein, C., Huntington, S., Kobberdahl, W., Leonard, W. and Zetocha, D. (1990) *Take Charge: Economic Development in Small Communities.* North Central Regional Center for Rural Development.

Baker, H.R. (1994) The Community Development Process, *Making Community Agri-Based Development Happen on the Prairies!* Saskatoon: Extension Division, University of Saskatchewan.

Baldwin, F. (1995) Enterprising communities: West Virginia's winning ways, *Appalachia* 28 (3-4), pp. 8-15.

Barnes, T. and Hayter, R. (1993) *Economic Restructuring and Local Development at the Margin: Forest Communities in Coastal British Columbia.* Thunder Bay, Ontario: Institute for Northern Studies, Lakehead University.

Baron, J. and Bielby, W. (1980) Bringing firms back in: stratification segmentation and the organization of work, *American Sociological Review* 45, pp. 737-765.
228

Baron, R.R. (1975) *Seasonality in Tourism.* London: Economist Intelligence Unit.

Bascom, J., Gordon, R., Cedillo, M., Scull, R. and King, T. (1998) Revisiting the 'rural revolution'. In: *East Carolina: Industrial Restructuring on the Coastal Plain Region*, manuscript in preparation.

Bateman, D. and Ray, C. (1994) Farm pluriactivity and rural policy: some evidence from Wales, *Journal of Rural Studies* 10, pp. 1-14.

Bateson, F.W. (ed) (1946) *Towards a Socialist Agriculture: Studies by a Group of Fabians.* London: Gollancz.

Beesley, K.B. (1981) The mature urban fringe. In: Beesley, K.B. and Russwurm, L.H. (eds) *The Rural-Urban Fringe: Canadian Perspectives.* Toronto: York University, Atkinson College, Geographical Monographs No. 10, pp. 87-95.

Beesley, K.B. (1983a) Citizen participation and the community, *The Operational Geographer* 2, pp. 45-47.

Beesley, K.B. (1983b) The quality of community life, *Geographical Perspectives* 51, pp. 54-57.

Beesley, K.B. (1985) *The Quality of Life in the Urban Field: Urbanites in the City's Countryside.* Waterloo: University of Waterloo, Department of Geography, unpublished Ph.D. thesis.

Beesley, K.B. (1988a) Living in the urban field. In: Coppack, P.M., Russwurm, L.H. and Bryant, C.R. (eds) *Essays on Canadian Urban Process and Form, Volume III: The Urban Field.* University of Waterloo, Department of Geography Publication Series No. 30, pp. 131-156.

Beesley, K.B. (1988b) *Perceptions of the Urban Fringe: A Comparative Assessment.* Toronto: York University, Department of Geography, Discussion Paper No. 35.

Beesley, K.B. (ed) (1991a) *Rural and Urban Fringe Studies in Canada.* Toronto: York University, Atkinson College, Geographical Monographs No. 21.

Beesley, K.B. (1991b) Rural and urban fringe studies in Canada: retrospect and prospect. In: Beesley, K.B. (ed) *Rural and Urban Fringe Studies in Canada.* Toronto: York University, Atkinson College, Geographical Monographs No. 21, pp. 1-42.

Beesley, K.B. (1993) *The Rural-Urban Fringe: A Bibliography.* Peterborough: Trent University, Department of Geography, Occasional Paper 15.

Beesley, K.B. (1994a) *Sustainable Development and the Rural-Urban Fringe: A Review of the Literature.* Winnipeg: University of Winnipeg, Institute of Urban Studies, Issues in Sustainability No. 3.

Beesley, K.B. (1994b) *Local Concerns in Urban Fringe Environments: A Case Study of Lakefield and Smith Township, Ontario*. Truro: Nova Scotia Agricultural College, Department of Humanities, Rural Studies Working Papers No. 5.

Beesley, K.B. (1994c) *Satisfaction with Life and Community in a Non-metropolitan Region: Peterborough, Ontario*. Truro: Nova Scotia Agricultural College, Department of Humanities, Rural Studies Working Papers No. 7.

Beesley, K.B. (1995a) *Modelling the Subjective Quality of Life in the Rural-Urban Fringe: Comparative Assessments in Southern Ontario*. Truro: Nova Scotia Agricultural College, Department of Humanities, Rural Studies Working Papers No. 15.

Beesley, K.B. (1995b) *Community Satisfaction in the Urban Field: An Ontario Example*. Truro: Nova Scotia Agricultural College, Department of Humanities, Rural Studies Working Papers No. 17.

Beesley, K.B. (1995c) *Satisfaction with Life and Community in Urban Field Environments: Statistical Findings from Selected Ontario Studies, 1984-1993*. Truro: Nova Scotia Agricultural College, Rural Research Centre, Research Paper No. 7.

Beesley, K.B. (1997) Metro-nonmetro comparisons of satisfaction in the rural-urban fringe: Southern Ontario, *Great Lakes Geographer* 4 (1), pp. 57-66.

Beesley, K.B. and Bowles, R.T. (1991) Change in the countryside: the turnaround, the community and the quality of life, *The Rural Sociologist* 11 (4), pp. 37-46.

Beesley, K.B. and Macintosh, P.J. (1993) *The Quality of Life in Urban Fringe Environments: Farm - Non-farm Comparisons in Southwestern Ontario*. Truro: Nova Scotia Agricultural College, Department of Humanities, Rural Studies Working Papers No. 4.

Beesley, K.B. and Macintosh, P.J. (1994) The quality of life in urban fringe environments: farm - non-farm comparisons in southwestern Ontario, *NESTVAL Proceedings* 23, pp. 75-87.

Beesley, K.B. and Macintosh, P.J. (1995a) *Housing Satisfaction in a Non-metropolitan City: the Case of Peterborough*. Truro: Nova Scotia Agricultural College, Rural Research Centre, Research Paper No. 6.

Beesley, K.B. and Macintosh, P.J. (1995b) Perceptions of crime and safety: Amherst and area. In: Beesley, K.B. and Macintosh, P.J. (eds) *Rural Research in the Humanities and Social Sciences II*. Truro: Nova Scotia Agricultural College, Rural Research Centre, Rural Research Centre Colloquia Proceedings, No. 2, pp. 47-68.

Beesley, K.B. and Macintosh, P.J. (1998) The quality of life in urban fringe environments: farm - non-farm comparisons in Southwestern Ontario, *Social Indicators Research* (in press).

Beesley, K.B. and Russwurm, L.H. (1989) Social indictors and quality of life research: towards synthesis, *Environments: A Journal of Interdisciplinary Research* 20, pp. 22-39.

Beesley, K.B. and Walker, G.E. (1990a) Residence paths and community perception: a case study from the Toronto urban field, *The Canadian Geographer* 34 (4), pp. 318-330.

Beesley, K.B. and Walker, G.E. (1990b) Local satisfactions and concerns in urban fringe areas, *Ontario Geography* 34, pp. 23-36.

Beesley, K.B., Bowles, R.T. and Johnston, C. (1993) *Life and Community Satisfaction in a Retirement Village: Bobcaygeon, Ontario*. Truro: Nova Scotia Agricultural College, Department of Humanities, Rural Studies Working Papers No. 3.

Beesley, K.B., Macintosh, P.J. and Bowles, R.T. (1995a) *Community Satisfaction in Non-metropolitan Areas: Seniors in City and Village Environments*. Truro: Nova Scotia Agricultural College, Department of Humanities, Rural Studies Working Papers No. 13.

Beesley, K.B., Macintosh, P.J. and Walker, G.E. (1995b) Community perception, community satisfaction and social interaction in the Toronto metropolitan fringe, *NESTVAL Proceedings* 24, pp. 8-34.

Beesley, K.B., Walker, G.E. and Everitt, J.C. (1996) Toward an understanding of the perception of rurality: Canadian perspectives, *NESTVAL Proceedings* 25, pp. 28-52.

Bell, C. and Newby, H. (1971) *Community Studies*. London: Allen and Unwin.

Blahna, D. (1991) Social bases for resource conflicts in areas of reverse migration. In: Lee, R.G., Field, D.R. and Burch, W.W. (eds) *Community and Forestry: Continuities in the Sociology of Natural Resources*. Boulder: Westview Press, pp. 159-178.

Blakely, E.J. (1994) *Theoretical Approaches for a Global Community, Community Development in Perspective*. Ames, Iowa: Iowa State University Press.

Blakeslee, S. (1997) Baby talk builds brains: foundations for rational thinking set by age one, study says, *The Ottawa Citizen* 17 April, p. A15.

Blank, U. (1989) *The Community Tourism Industry Imperative: The Necessity, The Opportunities, Its Potential*. State College, PA: Venture Publishing.

Bluestone, B. and Harrison, B. (1982) *The Deindustrialization of America.* New York: Basic Books.

Bluestone, B., Murphy, W.M. and Stevenson, M. (1973) *Low Wages and the Working Poor.* Ann Arbor: University of Michigan Press.

Bollman, R.D. (1994) A preliminary typology of rural Canada. In: Bryden, J.M. (ed) *Towards Sustainable Rural Communities: The Guelph Seminar Series.* Guelph: University School of Rural Planning and Development, University of Guelph, pp. 141-144.

Bollman, R.D., Fuller, T. and Ehrensaft, P. (1992) Rural jobs: trends and opportunities, *Canadian Journal of Agricultural Economics* 40, pp. 605-622.

Bovaird, A.G., Martin, S.J., Osborne, S., Tricker, M. and Waterson, P. (1992) *Managing Social and Community Development Programmes in Rural Areas: a Management Review of the Rural Development Commission's Social Programme.* Birmingham: Aston Business School.

Bovaird, A.G., Martin, S.J., Millward, A. and Tricker, M.J. (1995) *Evaluation of Rural Action for the Environment: Three Year Review of Effectiveness.* Birmingham: Aston Business School.

Bowler, I.R. (1985) Some consequences of the industrialization of agriculture in the European Community. In: Healey, M.J. and Ilbery, B.W. (eds) *The Industrialization of the Countryside.* Norwich: GeoBooks, pp. 75-98.

Bowler, I.R. and Lewis, G.J. (1991) Community involvement in rural development: the example of the Rural Development Commission. In: Champion, T. and Watkins, C. (eds) *People in the countryside: studies of social change in rural Britain.* London: Chapman, pp. 160-177.

Bowles, B.A. and Teale, R. (1994) Communications services in support of collaborative health care, *British Telecom Technology Journal* 12, pp. 29-44.

Bowles, R.T. and Beesley, K.B. (1991) Quality of life, migration to the countryside, and rural community growth. In: Beesley, K.B. (ed) *Rural and Urban Fringe Studies in Canada.* Toronto: York University, Atkinson College, Geographical Monographs No. 21, pp. 45-66.

Bridges, W. (1991) *Managing Transitions: Making the Most of Change.* Reading, Mass.: Addison-Wesley.

British Telecom (1994) *The BT Guide for People who are Disabled or Elderly - 1994.* London: British Telecommunications.

Britton, S. (1991) Tourism, capital, and place: towards a critical geography of tourism, *Environment and Planning D: Society and Space* 9, pp. 451-478.

Bryant, C.R. (1991) *Community Development and Restructuring of Rural Employment in Canada.* Paper presented at the Contemporary Social and

Economic Restructuring of Rural Areas International Seminar UK 12-17 August.

Bryant, C.R. (1994) *Sustainable Community Analysis Workbook 1: Working Together Through Community Participation, Co-operation and Partnerships.* St. Eugene, Ontario: Econotrends Limited.

Bryant, C.R. (1995) *Community Economic Development: Changing the Shape of the Future through the Power of the People.* Paper presented at the conference Preparing for Now! Community Tools and Strategies for Economic Renewal, Community Economic Development Centre, Simon Fraser University, Vancouver, British Columbia, 17 June.

Bryant, C.R., Russwurm, L.H. and McLellan, A.G. (1982) *The City's Countryside.* London: Longman.

Bryden, J.M., Black, J.S., Conway, E. and Shucksmith, D.M. (1994) *WISL LEADER Evaluation.* Stornaway: WISL LEADER Group.

Buchanan, J. (1983) *The Mobility of Disabled People in a Rural Environment.* London: Royal Association for Disability and Rehabilitation.

Bunce, M. (1981) Rural sentiment and the ambiguity of the urban fringe. In: Beesley, K.B. and Russwurm, L.H. (eds) *The Rural-Urban Fringe: Canadian Perspectives.* Toronto: York University, Atkinson College, Geographical Monographs No. 10, pp. 109-120.

Bunce, M. (1982) *Rural Settlement in an Urban World.* New York: St. Martin's Press.

Burns, D., Hambleton, R. and Hoggett, P. (1994) *The Politics of Decentralization: Revitalizing Local Democracy.* London: MacMillan.

Butler, M. *et al.* (1995) *The Coastal Zone Canada 1994.* Dartmouth, Nova Scotia: Conference Secretariat, Bedford Institute of Oceanography.

Caldock, K. and Wenger, C. (1992) Health and social service provision for elderly people: the need for a rural model. In: Gilg, A.W., Briggs, D., Dilley, R., Furuseth, O. and McDonald, G. (eds) *Progress in Rural Policy and Planning Vol. II.* London: Belhaven Press.

Campbell, A. (1981) *The Sense of Well-Being in America.* New York: McGraw-Hill.

Cassidy, S. (1994) Leadership and the management of change in community transport organizations. In: Smith, J. (ed) *Transport and Welfare.* Salford: University of Salford.

Chanard, A. (1994) *Diagnosis of the area and the mounting of a development project.* LEADER Co-ordinating Unit, AEIDL, Chaussee St.-Pierre 260.

Chouinard, V. (1997) Making space for disabling differences: challenging ableist geographies, *Environment and Planning D: Society and Space* 15, pp. 379-390.

Churchill, W.S. (1949) *The Second World War*. Volume 2. London: Cassell.

Clark, D. (1985) *Post-Industrial America.* New York: Methuen.

Clark, D. and Woollett, S. (1990) *English Village Services in the 80s*. London: Rural Development Commission.

Clark, G. (1982) The Agricultural Census - United Kingdom and United States, *Concepts and Methods in Modern Geography*, 35.

Cloke, P. (1993) On 'problems and solutions'. The reproduction of problems for rural communities in Britain during the 1980s, *Journal of Rural Studies* 9, pp. 113-121.

Cloke, P. and Goodwin, M. (1992) Conceptualizing countryside change: from post-Fordism to rural structured coherence, *Transactions of the Institute of British Geographers* 17, pp. 321-336.

Cloke, P. and Little, J. (1990) *The Rural State: Limits to Planning in Rural Society*. Oxford: Clarendon Press.

Cloke, P., Doel, M., Matless, D., Phillips, M. and Thrift, N. (1994) *Writing the Rural*. London: Paul Chapman.

Commission of the European Communities (1991a) *R and D on Telematic Systems for Rural Areas. Workplan 1991*. Brussels: Directorate General XIII.

Commission of the European Communities (1991b) *Notice to Member States (91/C 73/14).*

Commission of the European Communities (1995) *TELEMATICS for the Integration of Disabled and Elderly people (TIDE)*. Brussels: Directorate General XIII.

Committee on Land Utilization (1934) *Land Utilization in Minnesota: A State Program for the Cut-Over Lands*. Final Report of the Committee on Land Utilization, 4 August 1932. Minneapolis: The University of Minnesota Press.

Commons Select Committee for Agriculture (1993) *Effects of the Beer Orders on the Brewing Industry and Consumers*. London: HMSO.

Coppack, P.M., Beesley K.B. and Mitchell, C.J.A. (1990) Rural attractions and rural development: Elora, Ontario case study. In: Dykeman, F.W. and Sackville, N.B. (eds) *Entrepreneurial and Sustainable Rural Communities.* Mount Allision University, Rural and Small Town Research and Studies Programme, pp. 115-128.

Coppock, J.T. (1964) *An Agricultural Atlas of England and Wales*. London: Faber.

Corbin, D. (1981) *Life, Work and Rebellion in the Coal Fields: The Southern West Virginia Miners, 1880-1922*. Urbana and Chicago: The University of Illinois Press.

Countryside Commission (1990) *Local Countryside Action: Policies and Practice,* CCP 306. Cheltenham: Countryside Commission.

Couto, R. (1994) The future of the welfare state: the case of Appalachia. In: Obermiller, P. and Philliber, W. (eds) *Appalachia in an International Context: Cross-National Comparisons of Developing Regions*. Westport: Praeger, pp. 1-28.

Cox, G., Lowe, P. and Winter, M. (1986) Agriculture and conservation in Britain: a policy community under siege. In: Cox, G., Lowe, P. and Winter, M. (eds) *Agriculture, People and Policies*. London: Allen and Unwin, pp. 118-215.

Cox, K.R. and Mair, A. (1988) Locality and community in the politics of local economic development, *Annals of the Association of American Geographers* 78, pp. 307-325.

Crosbie, P. (1994) McDonald's serves up 3,000 jobs, *Daily Express* 29 December, p. 51.

Dahms, F.A. and Hallman, B. (1991) Population change, economic activity and amenity landscapes at the outer edge of the urban fringe. In: Beesley, K.B. (ed) *Rural and Urban Fringe Studies in Canada*. Toronto: York University, Atkinson College, Geographical Monographs No. 21, pp. 67-90.

Davidson, C. (1998) Drinking by numbers, *New Scientist* 11, pp. 36-39.

Davis, A. and Clifford, W. (1973) *Rural and Urban Population Change in North Carolina, 1960-1970*. Raleigh: North Carolina State University.

D'Costa, A. (1993) State-sponsored internationalization: restructuring and development of the steel industry. In: Graham, J. and Markusen, A. (eds) *Trading Industries, Trading Regions*. New York: Guilford Press, pp. 92-139.

Deavers, K. (1991) 1980s a decade of broad rural stress, *Rural Development Perspectives* 7 (3), pp. 2-5.

Debbage, K. (1990) Oligopoly and the resort cycle in the Bahamas, *Annals of Tourism Research* 17, pp. 513-527.

DeHart, J. (1990) Out of state garbage dumping in West Virginia, *Proceedings of the Fifth Conference on Appalachian Geography,* pp. 5-15.

Denbigh, A. (1996) British and Irish telecottages, *Teleworker* 3 (June-July), pp. 20-21.

Denter, E. (1994) British beer lovers' familiar pubs fast disappearing, *The Globe and Mail* 23 May, p. A6.

Department of Defence (1993) *1961-93, Civilian Re-use of Former Military Bases*. Washington, DC: Office of Economic Adjustment.

Department of the Environment (1990) *This Common Inheritance: Britain's Environmental Strategy.* London: HMSO. Cmnd 1200.

Department of the Environment and Welsh Office (1991) *The Control of Pollution (Silage, Slurry and Agricultural Fuel Oil) Regulations 1991.* London: HMSO. Statutory Instrument 324.

Department of the Environment (1992) *David Maclean launches Rural Action.* London: DoE Press Release 821.

District of Squamish (1994) *Tourism Development Plan.* Squamish, British Columbia.

Dixon, C. (1978) The changing structure of the British brewing industry, *Geography* 63, pp. 108-113.

Dolbearne, K.M. and Dolbearne, P. (1976) *American Ideologies: the Competing Political Beliefs of the 1970s.* Chicago: Rand McNally.

Doub, Jr., A. and Crabtree, L. (1973) *Tobacco in the United States.* Washington, DC: United States Department of Agriculture.

Durand, Jr., L. and Bird, E.T. (1950) The Burley Tobacco region of the Mountain South, *Economic Geography* 26, pp. 274-300.

The Economist (1991) They told you so, 31 August, pp. 23, 28.

Edwards, R., Reich, M. and Gordon, D. (eds) (1975) *Labor Market Segmentation.* Lexington: D. C. Heath.

ENDS (1990) *River Authority Wants Farm Waste Plans to Reduce Water Pollution.* ENDS Report 183. London: Environmental Data Services, pp. 6-7.

English Nature (1991) *Community Action for Wildlife.* Peterborough: English Nature.

English Nature (1993) *Strategy for the 1990.* Peterborough: English Nature.

Environment Agency (1996a) *The Environment of England and Wales: A Snapshot.* Bristol: Environment Agency.

Environment Agency (1996b) *Water Pollution Incidents in England and Wales 1995.* London: HMSO.

Environment Agency (1998) *The State of the Environment of England and Wales: Fresh Waters.* Bristol: Environment Agency.

Estall, R. (1968) Appalachian state: West Virginia as a case study in the Appalachian regional development problem, *Geography* 53 (238), pp. 1-24.

Ethos Research Associates (1995) *Rethinking Government 1994.* Ottawa: Ethos Research Associates.

Etzioni, A. (1993) *The Spirit of Community: Rights, Responsibilities, and the Communitarian Agenda.* New York: Crown Publishers, Inc.

Evans, N.J. and Ilbery, B.W. (1989) A conceptual framework for investigating farm-based accommodation and tourism in Britain, *Journal of Rural Studies* 5 (3), pp. 257-266.

Evans, S.B. (1997) Where now? Mountain empire's Burley Tobacco growers ponder their future, *Bristol Herald Courier* 21 September, pp. 1A, 4A.

Everitt, J.C. and Bowler, I.R. (1996) Bitter-sweet conversions: changing times for the British pub, *Journal of Popular Culture* 30 (2), pp. 101-122.

Filion, P. (1991) Local economic development as response to economic transition, *Canadian Journal of Regional Science* 14, pp. 347-370.

Fisher, J. and Mitchelson, R. (1981) Forces of change in the American settlement pattern, *Geographical Review* 71, pp. 289-310.

Fisk, M.J. (ed) (1989) *Alarm Systems and Elderly People.* London: The Planning Exchange.

Flora, J.L., Gale, E., Schmidt, F.E., Green, G.P. and Flora, C.B. (1993) From the grassroots, *Case Studies of Eight Rural Self-Development Efforts,* 59 (2). Agriculture and Rural Economy Division, Economic Research Service, US Department of Agriculture.

Fontan, J-M. (1993) *Community Economic Development Literature: A Critical Review of Canadian, American, and European.* Vernon, British Columbia: CCE/Westcoast Publications.

Fozard, J.L., Graafmans, J.A.M., Rietsema, J., Bovman, H. and van Bento, A. (1994) Ageing, health and technology. In: Ekberg, J. (ed) *Elderly, Elderly Disabled and Technology.* Brussels: Commission of the European Community DGV and DGXIII.

Freshwater, D., Hu, D., Wojan, T.R. and Goetz, S.J. (1996) *Testing for Effects of Federal Economic Development Agencies: Does TVA make a Difference?* Lexington: University of Kentucky, Department of Agricultural Economics, Staff Paper No. 362.

Frye, A. (1997) What the DDA will do for transport, *Disability Now* July, p. 22.

Gans, H. (1972) The positive functions of poverty, *American Journal of Sociology* 78, pp. 275-289.

Gant, R. (1995) Ageing and disability in the British countryside, *Norois* 43, pp. 339-354.

Gant, R. (1997) *Transport for Disabled People in the British Countryside*. Paper presented to the Third Franco-British Conference for Rural Geography, Nantes.

Gant, R. and Smith, J. (1991) The elderly and disabled in rural areas: travel patterns in the North Cotswolds. In: Champion, T. and Watkins, C. (eds) *People in the Countryside*. London: Paul Chapman Publishing.

Gant, R. and Walford, N. (1998) Telecommunications support for elderly and disabled people in rural Britain, *Health and Place* 4(3), pp. 245-263.

Garfield, K. (1994) Called to serve, *The Charlotte Observer* 25 December, pp. A1, A10.

Garkovich, L.E. (1989a) Local organizations and leadership in community development. In: *Community Development in Perspective*. Ames, Iowa: Iowa State University Press.

Garkovich, L.E. (1989b) *Population and Community in Rural America*. New York: Greenwood Press.

Gasson, R. (1988) Farm diversification and rural development, *Journal of Agricultural Economics* 39 (2), pp. 175-182.

Gates, J. (1993) Chain reaction to pub sell-off, *Daily Express* 7 June, p. 36.

Gaventa, J. (1987) The poverty of abundance revisited, *Appalachian Journal* 15 (1), pp. 24-33.

Gilg, A.W. and Battershill, M.R.J. (1997) Farmer reaction to agri-environment schemes: a study of participants in South-West England and the implications for research and policy development. In: Ilbery, B.W., Chiotti, Q. and Rickard, T. *Agricultural Restructuring and Sustainability*. Wallingford: CAB International.

Gill, A. and Reed, M. (1997) The re-imaging of a Canadian resource town: post-productivism in a North American context, *Applied Geographic Studies* 1, pp. 129-147.

Gillenwater, M. (1977) Mining settlements of southern West Virginia. In: Adkins, H., Ewing, S. and Zimolzak, C. (eds) *West Virginia and Appalachia*. Dubuque: Kendall-Hunt, pp. 132-157.

Glade, O. and Stillwell, D. (1986) *North Carolina: People and Environments*. Boone, North Carolina: GEO-APP Publishing Co.

Gleeson, B. (1996) A geography of disabled people? *Transactions of the Institute of British Geographers* 21, pp. 387-396.

Gleeson, B. (1997) Community care and disability, *Progress in Human Geography* 21, pp. 199-224.

Gloucestershire County Council (1986) *North Cotswolds Surveys: Parish Profiles.* Gloucester.

Gloucestershire County Council (1991) *Community Care Plan 1991-1994.* Gloucester.

Gofton, L.R. (1986) A woman's place? Pubs in the leisure society, *Food Marketing* 2, pp. 163-179.

Gold, G.L. (1975) *St. Pascal: Changing Leadership and Social Organization in a Québec Town.* Toronto: Holt, Rinehart and Winston.

Golledge, R.G. (1996) A response to Imrie, *Transactions of the Institute of British Geographers* 21, pp. 404-411.

Gordon, D. (1972) *Theories of Poverty and Underemployment.* Lexington: D.C. Heath.

Gordon, R. (1996) *Rural non-farm population growth in eastern North Carolina.* East Carolina University, unpublished M.A. thesis.

Gotsch-Thomson, S. (1988) Ideology and welfare reform under the Reagan administration. In: Tomaskovic-Devey, D. (ed) *Poverty and Social Welfare in the United States.* Boulder: Westview, pp. 222-249.

Government of Canada (1995) *Rural Canada: A Profile.* Ottawa: Government of Canada, Federal Interdepartmental Committee on Rural and Remote Canada, LM-347-02-95E.

Graham, S. (1991) Telecommunications and the local economy: some emerging policy issues, *Local Economy* 6, pp. 116-136.

Greenop, D. Pearson, I. and Johnson, T. (1994) Broadband - liberating the customer, *British Telecommunications Engineering* 12, pp. 252-258.

Griliches, Z. (1996) *Education, Human Capital and Growth: A Personal Perspective.* Cambridge: Harvard University, Discussion Paper Number 1745.

Grimes, S. (1992) Exploiting information and telecommunications technologies for rural development, *Journal of Rural Studies* 8, pp. 269-278.

The Guardian (1995) Protest groups unite to reclaim their lost land, 24 April.

Haines-Young, R. and Watkins, C. (1996) The rural data infrastructure, *International Journal Geographical Information Systems* 10 (1), pp. 21-46.

Hajesz, D. and Dawe, S.P. (1997) De-mythologizing rural youth exodus. In: Bollman, R.D. and Bryden, J.M. (eds) *Rural Employment: An International Perspective.* Brandon, Brandon University: Rural Development Institute for the Canadian Rural Restructuring Foundation.

Halfacree, K. (1993) Locality and social representation: space, discourse and alternative definitions of the rural, *Journal of Rural Studies* 9, pp. 23-37.

Halfacree, K. (1995) Talking about rurality: social representations of the rural as expressed by residents of six English parishes, *Journal of Rural Studies* 11, pp. 1-20.

Halfacree, K. (1996) Out of place in the country: travellers and the rural idyll, *Antipode* 28, pp. 42-71.

Halseth, G. (1997) Community economic development groups in non-metropolitan British Columbia: interpreting the representativeness of civil society organizations, *Small Town* 28 (1), pp. 18-23.

Hannigan, J. (1995) *Environmental Sociology: A Social Constructionist Perspective*. London: Routledge.

Harper, S. (1989) The British rural community: an overview of perspectives, *Journal of Rural Studies* 5, pp. 161-184.

Hart, J.F. (1998) *The Rural Landscape*. Baltimore: The Johns Hopkins University Press.

Hart, J.F. and Chestang, E.L. (1978) Rural revolution in East Carolina, *Geographical Review* 68, pp. 435-458.

Hart, J.F. and Chestang, E.L. (1996) Turmoil in tobaccoland, *Geographical Review* 86, pp. 550-572.

Hart, J.F. and Morgan, J.T. (1995) Spersopolis, *Southeastern Geographer* 35, pp. 103-117.

Harris, B. (1995) Telecentre challenge delayed, *Gloucestershire Rural Voice* April, p. 8.

Harvey, D. (1985) *The Urbanization of Capital*. Oxford: Blackwell.

Harvey, D. (1987) Flexible accumulation through urbanization: reflections on post-modernism in the American city, *Antipode* 19, pp. 260-286.

Hawkins, E.A., Bryden, J.M., Gilliatt, N. and MacKinnon, N. (1993) Engagement in agriculture 1987-1991: a West European perspective, *Journal of Rural Studies* 9 (3), pp. 277-290.

Hawkins, K. (1984) *Environment and Enforcement: Regulation and the Social Definition of Pollution*. Oxford: Clarendon Press.

Hawkins, K. and Pass, C.L. (1979) *The Brewing Industry: a Study in Industrial Organization and Policy*. London: Heinemann.

Hawkins, L. (1995) *Mapping the Diversity of Rural Economies: A Preliminary Typology of Rural Canada*. Ottawa: Statistics Canada, Agriculture Division, Working Paper No. 29.

Hawkins, L. and Bollman, R.D. (1994) Revisiting rural Canada - it's not all the same. In: Statistics Canada, *Canadian Agriculture at a Glance*. Ottawa: Statistics Canada, Cat. No. 96-301, pp. 78-80.

Headwaters Regional Development Commission (1992) *Overall Economic Development Program Report*.

Heap, P., Baker, D. and Hermon, J. (1993) Visions for villages, *Leicester Mercury* 4 June, pp. 24-25.

Henwood, M. (1992) *Through a Glass Darkly: Community Care and Elderly People*. Poole: King's Fund Institute.

The Herald Dispatch (1990) Here we go again - let's stomp on West Virginia, 29 March, pp. A1-A2

Hertzman, C. (1994) *The Lifelong Impact of Childhood Experiences: A Population Health Perspective*. Toronto: The Canadian Institute for Advanced Research, Working Paper No. 47.

Hill, C. (1972) *The World Turned Upside Down*. London: Temple Smith.

HMSO (1967) *A Type of Farming Map for England and Wales*. Soil Survey of England and Wales.

HMSO (1992) *Community Care: Managing the Cascade of Change*. London: HMSO.

HMSO (1995) *Rural England: A Nation Committed to a Living Countryside,* Cmnd. 3016. London: HMSO.

HMSO (1998) *Modernizing Local Government: Improving Local Services through Best Value*. London: HMSO.

Hoffman, K. and Dupont, J-M. (1992) *Community Health Centres and Community Development*. Ottawa: Health Services and Promotion Branch, Health and Welfare Canada.

Holtkamp, J., *et al.* (1997) Economic development effectiveness of multicommunity development organizations, *Journal of the Community Development Society* 28 (2), pp. 242-256.

House of Commons Committee of Public Accounts (1991) *Advisory Services to Agriculture*. HC Paper 465, Session 1990-91. London: HMSO.

House of Commons Environment Committee (1987) *Pollution of Rivers and Estuaries*. HC Paper 183-1, Third Report, Session 1986-7. London: HMSO.

Howkins, A. (1989) The ploughman's rest', *New Statesman and Society* 14 April, p. 11.

Humphery, D. (1992) *Dying With Dignity: What You Need to Know About Euthanasia*. New York: St. Martin's.

Ilbery, B.W. (1981) Dorset agriculture: a classification of regional types, *Transactions of the Institute of British Geographers NS* 6, pp. 214-227.

Ilbery, B.W. and Bowler, I.R. (1998) From agricultural productivism to post-productivism. In: Ilbery, B.W. (ed) *The Geography of Rural Change*. Harlow: Longman.

Imrie, R. (1996) Ableist geographers, disableist spaces: towards a reconstruction of Golledge's 'Geography and the disabled', *Transactions of the Institute of British Geographers* 21, pp. 397-403.

Imrie, R. (1997) Challenging disabled access in the built environment, *Town Planning Review* 68, pp. 423-448.

The Independent on Sunday (1995) The peaceful revolutionary behind 'The Land is Ours', 30 April.

Innskeep, E. (1991) *Tourism Planning: An Integrated and Sustainable Development Approach*. New York: Van Nostrand Reinhold.

Ioannides, D. (1995) Strengthening the ties between tourism and economic geography: a theoretical agenda, *Professional Geographer* 47, pp. 49-60.

Ioannides, D. and Debbage, K. (1997) Post-Fordism and flexibility: the travel industry polyglot, *Tourism Management* 18, pp. 229-241.

Jakle, J. and Wilson, D. (1992) *Derelict Landscapes: The Wasting of America's Built Environment*. Rowman and Littlefield, Savage.

Jessop, B. (1990) Regulation theories in retrospect and prospect, *Economy and Society* 19, pp. 153-216.

Jobes, P.C., Stinner, W.F. and Wardwell, J.M. (1992) *Community, Society and Migration: Non-economic Migration in America*. Lanham, Maryland: University Press of America.

Johnson, K. (1985) *The Impact of Population Change on Business Activity in Rural America*. Boulder: Westview Press.

Kain, R.J.P. and Prince, H.C. (1985) *The Tithe Surveys of England and Wales*. Cambridge, Cambridge University Press.

Kalleberg, A., Wallace, M. and Althauser, R. (1981) Economic segmentation, worker power and income inequality, *American Journal of Sociology* 87, pp. 651-683.

Katovich, M. and Reese, W. (1987) The regular: full-time identities and membership in an urban bar, *Journal of Contemporary Ethnography* 16, pp. 308-343.

Keane, M.J. (1990) Economic development capacity amongst small rural communities, *Journal of Rural Studies* 6, pp. 291-301.

Keary, T.J., Fredman, L., Taler, G.A., Datta, S., Levenson, S.A. (1994) *Indicators of Quality Medical Care for the Terminally Ill in Nursing Homes*, American Geriatrics Society, Special Article.

Keating, D. and Mustard, J.F. (1993) *The National Forum on Family Security - Social Economic Factors and Human Development*. Toronto: The Canadian Institute for Advanced Research.

Killian, M.S. and Beaulieu, L.J. (1995) Current status of human capital in the rural US. In: Beaulieu, L.J. and Mulkey, D. (eds) *Investing in People: The Human Capital Needs of Rural America*. Boulder: Westview Press, pp. 23-46.

Killian, M.S. and Parker, T.S. (1991) Education and local employment growth in a changing economy. In: US Department of Agriculture, *Education and Rural Economic Development: Strategies for the 1990s*. Washington, DC: United States Department of Agriculture, Economic Research Service, Report No. AGES-9153, pp. 93-113.

Knipscheer, K. (1994) Ageing populations, living conditions and health care policy. In: Ekberg, J. (ed) *Elderly, Elderly Disabled and Technology*. Brussels: Commission of the European Community DGV and DGXIII.

Knowland, T. (1993) *Changing the Guard? Institutional Change in Water Pollution Control*. University of East Anglia, unpublished Ph.D. thesis.

Kretzmann, J. and McKnight, J. (1993) *Building Communities from the Inside Out*. Evanston, Illinois: Center for Urban Affairs and Policy Research, Northwestern University.

Kusmin, L.D. (1994) *Factors Associated with the Growth of Local and Regional Economies: A Review of Selected Empirical Literature*. Washington, DC: United States Department of Agriculture, Economic Research Service, Staff Report No. AGES-9405.

Kusmin, L.D., Redman, J.M. and Sears, D.W. (1996) *Factors Associated with Rural Economic Growth: Lessons from the 1980s*. Washington, DC: United States Department of Agriculture, Economic Research Service, Technical Bulletin No. 1850.

Land Management Information Center (1995a) *1983 Public Ownership by County*.

Land Management Information Center (1995b) *Generalized Soil Atlas Peat and Bogs*.

Lefebvre, H. (1991) *The Production of Space*. Oxford: Blackwell.

Lewis, G.J. (1986) Welsh rural community studies: retrospect and prospect, *Cambria* 13, pp. 27-40.

Lewis, H. (1991) Fatalism or the coal industry? *Mountain Life and Work* 46, pp. 4-15. Reprinted in: Ergood, B. and Kuhre, B. (eds) *Appalachia: Social Context Past and Present*. Dubuque: Kendall-Hunt, pp. 221-229.

Liberty (1993) *The Road to Nowhere?* Report 93/4. London: Liberty.

Lievesley, K. and Warwick, M. (1992) *1991 Survey of Rural Services*. London: Rural Development Commission.

Local Government Management Board (1994) *Earth Summit Rio 1992: Agenda 21- a Guide for Local Authorities in the UK*. London: Local Government Management Board.

Long, L. (1982) Repopulating the countryside: a 1980 census trend, *Science* 217, pp. 1111-1116.

Lonsdale, R. (1966) Two North Carolina commuting patterns, *Economic Geography* 42, pp. 114-138.

Lonsdale, R. and Browning, C. (1971) Rural-urban locational preferences of southern manufacturers, *Annals of the Association of American Geographers* 68, pp. 255-268.

Looker, E.D. (1997) Rural-urban differences in youth transition to adulthood. In: Bollman, R.D. and Bryden, J.M. (eds) *Rural Employment: An International Perspective*. Brandon: Brandon University, Rural Development Institute for the Canadian Rural Restructuring Foundation.

Lowe, P. and Flynn, A. (1989) Environmental politics and policy in the 1980s. In: Mohan, J. (ed) *The Political Geography of Contemporary Britain*. London: MacMillan, pp. 225-279.

Lowe, P., Clark, J., Seymour, S. and Ward, N. (1992) *Pollution Control on Dairy Farms: An Evaluation of Current Policy and Practice*. London: SAFE Alliance.

Lowe, P., Clark, J., Seymour, S. and Ward, N. (1997) *Moralizing the Environment: Countryside Change, Farming and Pollution*. London: UCL Press.

Lowry, I. (1966) *Migration and Metropolitan Growth: Two Analytical Models*. San Francisco: Chandler Publishing.

Lundy, K.L.P. and Warme, B.D. (1986) *Sociology: A Window on the World*. Toronto: Methuen.

Luther, V. and Wall, M. (1986) *The Entrepreneurial Community: A Strategic Planning Approach to Community Survival*. Lincoln, Nebraska: Visions From the Heartland (Nebraska as Leader).

Magnusson, L. (1932) Company housing in the bituminous coalfields, *Monthly Labor Review* 10, pp. 215-222.

Maloney, W. and Richardson, J. (1994) Water policy-making in England and Wales: policy communities under pressure, *Environmental Politics* 3, pp. 110-138.

Mann, C.K. (1975) *Tobacco: The Ants and the Elephants.* Salt Lake City: Olympus Publishing Company.

Markusen, A. and Carlson, V. (1989) Deindustrialization in the American Midwest: causes and responses. In: Rodwin, L. and Sazanami, H. (eds) *Deindustrialization and Regional Economic Transformation: The Experience of the United States*. Boston: Unwin Hyman, pp. 29-59.

Marsden, T. and Flynn, A. (1993) Servicing the city: contested transitions in the rural realm, *Journal of Rural Studies* 9, pp. 201-214.

Marsden, T., Munton, R.J.C., Whatmore, S. and Flynn, A. (1986) Towards a political economy of agriculture: a British perspective, *International Journal of Urban and Rural research* 1, pp. 498-521.

Marsden, T., Murdoch, J., Lowe, P., Munton, R.J.C. and Flynn, A. (1993) *Constructing the Countryside*. London: UCL Press.

Marsh, D. and Rhodes, R.A.W. (eds) (1992). *Policy Networks in the British Government*. Oxford: Oxford University Press.

Martin, E. (1987) Almost heaven: the road south, *The Charlotte Observer* 27 March, pp. 1A; 14A.

Martin, S.J. (1995) Partnerships for local environmental action: observations on the first two years of Rural Action for the Environment, *Journal of Environmental Planning and Management* 38, pp. 149-166.

Martin, S.J. and Geddes, M.N. (1998) *Ex post Evaluation of the LEADER Initiative in England and Wales*. University of Warwick, Warwick Business School.

Martin, S.J., Tricker, M. and Bovaird, A.G. (1990) Rural development programmes in theory and practice, *Regional Studies* 24, pp. 268-276.

Martin, S.J., Bovaird, A.G., Green, J., Millward, A. and Tricker, M. (1994) *Rural Action for the Environment: Summary Evaluation of Progress in Selected 'Pacemaker networks'*. Cirencester: Action with Communities in Rural England.

Matless, D. (1992) Regional surveys and local knowledges: the geographical imagination in Britain, 1918-39, *Transactions of the Institute of British Geographers* 17, pp. 448-463.

Matless, D. (1998) *Landscape and Englishness.* London: Reaktion.

Matson, K.W. (1995) *A Brief History of Land Ownership Pattern on the Chippewa National Forest.* United States Department of Agriculture Forest Service, Chippewa National Forest, Deer River Ranger District, unpublished summary, June 29.

McCormick, J. (1991) *British Politics and the Environment.* London: Earthscan.

McGarry, M. (1985) *Community Alarm Systems for Older People.* Age Concern Scotland.

McGranahan, D.A. (1991) Introduction. In: US Department of Agriculture, *Education and Rural Economic Development: Strategies for the 1990s.* Washington, DC: United States Department of Agriculture, Economic Research Service, Report No. AGES-9153.

McGranahan, D.A. and Ghelfi, L.M. (1991) The education crisis and rural stagnation in the 1980s. In: US Department of Agriculture, *Education and Rural Economic Development: Strategies for the 1990s.* Washington, DC: United States Department of Agriculture, Economic Research Service, Report No. AGES-9153.

McGranahan, D. and Kassel, K. (1996) Education and regional employment in the 1980s: comparisons among OECD member countries. In: Bollman, R.D. and Bryden, J.M. (eds) *Rural Employment: An International Perspective.* Brandon: Brandon University, Rural Development Institute for the Canadian Rural Restructuring Foundation.

Merrifield, A. (1993) Place and space: a Lefebvrian reconciliation, *Transactions of the Institute of British Geographers* 18, pp. 516-531.

von Meyer, H. (1997) Rural employment in OECD countries: structure and dynamics of regional labour markets. In: Bollman, R.D. and Bryden, J.M. (eds) *Rural Employment: An International Perspective.* Brandon: Brandon University, Rural Development Institute for the Canadian Rural Restructuring Foundation.

von Meyer, H. and Muheim, P. (1997) Employment is a territorial issue, *The OECD Observer* 203 (January), pp. 22-26.

Milbourne, P. (ed) (1997) *Revealing Rural Others.* London: Pinter.

Miller, E. (1978) Mining and economic revitalization of the bituminous coal region of Appalachia, *Southeastern Geographer* 18 (2), pp. 81-92.

Millward, A. (1995) *An Evaluation of the Community Action for Wildlife Scheme: Final Report to English Nature*. Birmingham: Alison Millward Associates.

Milne, S. and Tufts, S. (1993) Industrial restructuring and the future of the small firm: the case of Canadian micro-breweries, *Environment and Planning A* 25, pp. 847-861.

Ministry of Agriculture and Fisheries (1946) *National Farm Survey of England and Wales (1941-43): A summary report*. London: HMSO.

Ministry of Agriculture, Fisheries and Food (1992) *Pilot Study to Cut Farm Waste Pollution*. MAFF News Release, 21 January. London: MAFF.

Ministry of Agriculture, Fisheries and Food (1997) *The Digest of Agricultural Statistics*. London: The Stationery Office

Ministry of Agriculture, Fisheries and Food and Welsh Office Agriculture Department (1991) *Code of Good Agricultural Practice for the Protection of Water*. London: MAFF.

Monbiot, G. (1995) Whose land? *The Guardian* 22 February.

Monckton, H. (1969) *A History of the English Public House*. London: Bodley Head.

Monopolies and Mergers Commission (MMC) (1989) *The Supply of Beer*. London: HMSO.

Moore, T. (1991) Eastern Kentucky as a model of Appalachia: the role of literary images, *Southeastern Geographer* 31 (2), pp. 75-89.

Moore, T. (1994) Core-periphery models, regional planning theory and Appalachian development, *Professional Geographer* 46 (3), pp. 316-331.

Moore, T. (1998) A southern West Virginia mining community revisited, *Southeastern Geographer* 38 (1), pp. 1-21.

Morgan, J.T. (1978) The ordering pit: a relict feature of the flue-cured tobacco landscape, *Southeastern Geographer* 18, pp. 102-114.

Morgan, J.T. (1996*) Farmer Response to Tobacco Policy in Southwestern Virginia*. Paper presented at Eighth Conference on Appalachian Geography, Pipestem State Park, WV, April.

Moseley, M. (1996) *The LEADER Programme 1992-1994: an Interim Assessment of a European Area-based Rural Development Programme*. Cheltenham: Countryside and Community Research Unit.

Moss, L. (1991) *The Government Social Survey: a History*. London: HMSO.

Muegge, J. and Ross, N. (1993) *Effective Community Decision-Making: Factsheet.* Ontario: Ministry of Agriculture and Food.

Mullens Advocate (1990) West Virginia mining towns vanishing as mining fades, 24 June.

Mundy, K. and Purcell, W. (1996) Both sides of the coin: an uncommon meeting about tobacco, *Horizons* 8 (5), pp. 1-4.

Munn, R. (1977) The development of model towns in the bituminous coalfields, *West Virginia History* 40 (3), pp. 243-253.

Murdoch, J. and Marsden, T. (1994) *Reconstituting Rurality.* London: UCL Press.

Murdoch, J. and Ward, N. (1997) Government and territoriality: the statistical manufacture of Britain's 'national farm', *Political Geography* 16, pp. 307-324.

Murphy, P.E. (1985) *Tourism a Community Approach.* London: Routledge.

Murphy, R. (1933) A southern West Virginia mining community, *Economic Geography* 9, pp. 51-59.

Murray, K.A.H. (1955) *Agriculture.* London: HMSO.

Mustard, J.F. (1994) *Health and Social Capital - A North American Perspective.* Toronto: The Canadian Institute for Advanced Research.

Nash, J.M. (1997) Fertile minds: how a child's brain develops, and what it means for child care and education, *Time* 9 June, pp. 47-54.

National Audit Office (1995) *National Rivers Authority: River Pollution from Farms in England - Report by the Comptroller and Auditor General.* HC 235, Session 1994-5. London: HMSO.

Nebraska Department of Economic Development., (1994) *The Nebraska Community Action Handbook: A Guide for Local Economic Development.* Nebraska Department of Economic Development.

Nelson Community Project (1994) *Discovering Your Community: A Co-operative Process for Planning Sustainability.* Victoria, BC: Harmony Foundation of Canada.

Newby, H. (1979) *Green and Pleasant Land?* London: Hutchinson.

Newby, H. (1987) *Country Life.* London: Weidenfeld and Nicolson.

Newman, M. (1972) *The Political Economy of Appalachia: A Case Study in Regional Integration.* Lexington: Lexington Books.

Newman, R.J. and Sullivan, D.H. (1980) Econometric analysis of business tax impacts on industrial location. What do we know, and how do we know it? *Journal of Urban Economics* 23, pp. 215-234.

NRA (1989a) *Guardians of the Water Environment*. London: NRA.

NRA (1989b) *Fact Sheet 13: Conservation*. London: NRA.

NRA (1990a) *Annual Report and Accounts 1989/90*. London: NRA.

NRA (1990b) *The Water Guardians, 7*. Bristol: NRA.

NRA (1990c) *Corporate Plan 1990/1991*. London: NRA.

NRA (1992a) *Annual Report and Accounts 1991/2*. Bristol: NRA.

NRA (1992b) *The Influence of Agriculture on the Quality of Natural Waters in England and Wales*. Water Quality Series No. 6. Bristol: NRA.

NRA (1992c) *Water Pollution Incidents in England and Wales - 1990*. Water Quality Series No. 7. Bristol: NRA.

NRA (1992d) *Water Pollution Incidents in England and Wales - 1991*. Water Quality Series No. 9. Bristol: NRA.

NRA (1993) *Water Pollution Incidents in England and Wales - 1992*. Water Quality Series No. 13. Bristol: NRA.

NRA (1994a) *Water Pollution Incidents in England and Wales - 1993*. Bristol: NRA.

NRA (1994b) *The Quality of Rivers and Canals in England and Wales, 1990-1992*. Bristol: NRA.

NRA (1995a) *Water Pollution Incidents in England and Wales - 1994*. Bristol: NRA.

NRA (1995b) *Drop in Major Pollution Incidents*. Bristol: NRA News Release 24.7.95.

NRA and MAFF (1990) *Water Pollution from Farm Waste 1989 (England and Wales)*. London: NRA.

NRA, South West Region (1992) *NRA in New Year Pollution Blitz*. NRA South West News Release, 8 January. Exeter: NRA South West.

Nutley, S.D. (1988) Unconventional modes of transport in rural Britain, *Journal of Rural Studies* 4, pp. 73-86.

OECD (1996) *Territorial Indicators of Employment: Focusing on Rural Development*. Paris: OECD.

O'Riordan, T. (1981) *Environmentalism*. London: Pion.

Osberg, L. (1995) The equity/efficiency trade-off in retrospect, *Canadian Business Economics* 3 (3), pp. 5-19.

Osborne, D. and Gaebler, T. (ed) (1992) *Reinventing Government: How the Entrepreneurial Spirit is Transforming the Public Sector*. New York: Penguin Books.

OPCS (1989) *The Prevalence of Disability Among Adults: OPCS Surveys of Disability in Great Britain Report No. 1*. London: HMSO.

Pacione, M. (1980) Differential quality of life in a metropolitan village, *Transactions of the Institute of British Geographers* 5, pp. 185-206.

Pacione, M. (1982) The use of objective and subjective measures of life quality in human geography, *Progress in Human Geography* 6, pp. 495-514.

Parr, H. (1998) Mental health, ethnography and the body, *Area* 30, pp. 28-37.

Perrin, R. and Sappie, G. (1990) *Economic Information Report, No. 83*. Department of Agriculture and Resource Economies, Raleigh, NC: North Carolina State University.

Phillips, M. (1998) Social perspectives. In: Ilbery, B.W. (ed) *The Geography of Rural Change*. London: Longman.

Phillips, P. and Brunn, S. (1978) Slow growth: a new epoch of American metropolitan evolution, *Geographical Review* 68, pp. 274-292.

Phillips, P.D. (1990) *Economic Development for Small Communities and Rural Areas*. Urbana and Chicago: University of Illinois Press.

Phillips, W.J. and Williams, H.S. (1991) The providers vs. the enabler, *Innovating*.

Philo, C. (1992) Neglected rural geographies: a review, *Journal of Rural Studies* 8, pp. 193-207.

The Pioneer (1994) Fish farm struggles with environmental problems. Bemidji, Minnesota, 10 October.

Pocahontas Fuel Co. (1936) *Drainage Tunnel of Pocahontas Fuel Company Incorporated*. New York: Pocahontas Fuel Company, Incorporated.

Pojul, D. (1994) *Exploiting Local Agricultural Resources*. Pierre 260, B-1040, Brussels.

Popkin, B.M. (1972) Economic benefits from the elimination of hunger in America, *Public Policy* Winter, pp. 133-153.

Pulver, G.C. (1993) *Building Your Community's Future (Satellite Programs on Community Economic Development)/Creating An Action Agenda: Seven Steps to Building an Action Agenda*. Urbana and Chicago: University of Illinois Press.

Raitz, K. and Ulack, R. (1984) *Appalachia, A Regional Geography: Land, People and Development*. Boulder: Westview Press.

Rawstorne, P. (1991) Where independence is the inn-thing, *Financial Times* 22 January, p. 13.

Reed, M. (1997) Power relations and community-based tourism planning, *Annals of Tourism Research* 24, pp. 566-591.

Reed, M. and Gill, A. (1997) Tourism, recreational, and amenity values in land allocation: an analysis of institutional arrangements in the post-productivist era, *Environment and Planning A* 29, pp. 2019-2040.

Reich, R.B. (1991) *The Work of Nations: Preparing Ourselves for 21st Century Capitalism*. New York: Vintage Books.

Relph, E. (1976) *Place and Placelessness*. London: Pion.

Richardson, I.E.G. and Riley, M.J. (1994) SAVIOUR - telemedicine in action, *Teleworking and Teleconferencing,* Colloquium of the Institute of Electrical Engineering, 7 June, 1994.

Robinson, G.M. (1990) *Agricultural Change*. Edinburgh: North British Publishing.

Robinson, G.M. (1994) The greening of agricultural policy: Scotland's Environmentally Sensitive Areas (ESAs), *Journal of Environmental Planning and Management* 37, pp. 215-225.

Robinson, G.M. (1997) Greening and globalizing: agriculture in the 'New Times'. In: Ilbery, B., Chiotti, Q. and Rickhard, T. (eds) *Agricultural Restructuring and Sustainability: A Geographical Perspective*. Wallingford: CAB International.

Robinson, M. (1995) Towards a new paradigm of community development, *Community Development Journal* 30 (1), pp. 21-30.

Rodwin, L. (1989) Deindustrialization and regional economic transformation. In: Rodwin, L. and Sazanami, H. (eds) *Deindustrialization and Regional Economic Transformation: The Experience of the United States*. Boston: Unwin Hyman, pp. 3-25.

Rogers, B. (1989) *Men Only*. London: Pandora Press.

Rost, J. (1991) *Leadership for the Twenty-First Century*. New York: Praeger Publishers.

Rural Action National Development Team (1996) *Bulletin for Network Members* (Autumn). Cirencester: National Development Team.

Rural Action National Steering Group (1993) *Rural Action for the Environment: an Introduction*. London: RDC.

Rural and Small Town Programme (1994) *Stepping Forward: Discovering Community Potential, Acting on Challenges*. Sackville, New Brunswick: Rural and Small Town Programme, Mount Allison University.

Rural Development Commission/OFTEL (1989) *Telecommunications in Rural England. Rural Development Commission Research Report No. 2*. London.

Rural Sociological Society Task Force on Persistent Rural Poverty (RSSTF) (1993) *Persistent Poverty in Rural America*. Boulder: Westview Press.

Rutherford, R.S.G. and Bateson, F.W. (1946) Co-operation in agriculture. In: Bateson, F.W. (ed) *Towards a Socialist Agriculture: Studies by a Group of Fabians*. London: Gollancz.

Ryan, W. (1976) *Blaming the Victim*. New York: Random House.

Rycroft, S. and Cosgrove, D. (1995) Mapping the modern state: Dudley Stamp and the Land Utilization Survey, *History Workshop Journal* 40, pp. 91-105.

Sack, R. (1986) *Human Territoriality*. Cambridge: Cambridge University Press.

Salamon, S. (1989) What makes rural communities tick? *Rural Development Perspectives* June, pp. 19-24.

Savage, M., Dicken, P. and Fielding, M. (1992) *Property, Bureaucracy and Culture: Middle-Class Formation in Contemporary Britain*. Routledge: London.

Sayer, A. (1989) The 'new' regional geography and problems of narrative, *Society and Space* 7, pp. 253-276.

Schultz, T.W. (1975) The value of the ability to deal with disequilibria, *Journal of Economic Literature* 13 (September), pp. 827-846.

Sears, D.W. and Reid, N.J. (1995) *Rural Development Strategies*. Chicago: Nelson-Hall.

Sears, D.W., Redman, J.M., Kusmin, L.D. and Killian, M.S. (1992) *Growth and Stability of Rural Economies in the 1980s: Differences Among Counties*. Washington, DC: United States Department of Agriculture, Economic Research Service, Staff Report AGES-9230.

Seymour, S., Cox, G. and Lowe, P. (1992) Nitrates in water: the politics of the Polluter-Pays-Principle, *Sociologia Ruralis* 32, pp. 82-103.

Seymour, S., Lowe, P., Ward, N. and Clark, J. (1997) Environmental 'others' and 'elites': rural pollution and changing power relations in the countryside. In: Milbourne, P. (ed) *Revealing Rural 'Others': Representation, Power and Identity in the British Countryside*. London and Washington: Pinter, pp. 57-74.

Shaffer, R. (1989) *Community Economics: Economic Structure and Change in Smaller Communities*. Ames, Iowa: Iowa State University Press.

Sherwood, K.B. and Lewis, G.R. (1994) Local appraisals and rural planning in England: an evaluation of current experience, *Progress in Rural Policy and Planning* 4, pp. 79-88.

Shoard, M. (1980) *The Theft of the Countryside*. London: Temple Smith.

Short, B. (1989) *The Geography of England and Wales in 1910: an Evaluation of Lloyd George's 'Domesday' of Landownership*. Cheltenham: HGRG Research Series No 22.

Short, B. (1997) *Land and Society in Edwardian Britain*. Cambridge: Cambridge University Press.

Short, B. and Watkins, C. (1994) The National Farm Survey of England and Wales 1941-3, *Area* 26 (3), pp. 288-293.

Short, B., Watkins, C., Foot, W. and Kinsman P. (forthcoming 1999) *State Surveillance and the Countryside in England and Wales, 1939-45*. Wallingford: CAB International.

Short, J. (1991) *Imagined Country*. London: Routledge.

Shragge, E. (1993) *Community Economic Development: In Search of Empowerment*. Montreal: Black Rose Books.

Shucksmith, M. (1993) Farm household behaviour and the transition to post-productivism, *Journal of Agricultural Economics* 44 (3), pp. 466-478.

Sibley, D. (1994) The sin of transgression, *Area* 26, pp. 300-303.

Sibley, D. (1995) *Geographies of Exclusion*. London: Routledge.

Simons, M. (1994) Era of the local bistro fast disappearing in France, *The Globe and Mail* 27 June, p. A8.

Skeffington, A. (1969) *People and Planning*. London: HMSO.

Skinner, A. (1992) *North Carolina Rural Profile: Economic and Social Trends Affecting Rural North Carolina*. Raleigh, NC: The North Carolina Rural Economic Development Center.

Smith, M.J. (1990) *The Politics of Agricultural Support in Britain: the Development of the Agricultural Policy Community*. Aldershot: Dartmouth Press.

Smith, N. (1984) *Uneven Development*. Oxford: Blackwell.

Stamp, L.D. (1947) *The Land of Britain: Its Use and Misuse*. London: Longman.

Stapledon, G. (1939) *The Plough-up Policy and Ley-farming*. London: Faber and Faber.

Strickland, J.L. (1993) *Place Meaning and Social Policy: the Settlement Movement in Eastern Kentucky, 1880-1914*. Charlotte: The University of North Carolina at Charlotte, unpublished Master's thesis.

Tennis, J. (1997) Survival skills, *Bristol Herald Courier* October 12, pp. 4D, 8D.

The Land is Ours (1995) *Newsletter* 1.

Thornton, P. (1993) Communications technology - empowerment or disempowerment, *Disability, Handicap and Society* 8, pp. 329-349.

Thorpe, F. and Pronay, N. (1980) *British Official Films in the Second World War*. Clio Press.

Thrift, N. (1987) Introduction: the geography of late twentieth-century class formation. In: Thrift, N. and Williams, P. (eds) *Class and Space*. London: Routledge and Kegan Paul, pp. 207-253.

Thrift, N. (1989) Images of social change. In: Hamnett, C., McDowell, L. and Sarre, P. (eds) *The Changing Social Structure*. London: Sage/Open University Press, pp. 12-42.

Tinker, A. (1989) *The Telecommunication Needs of Disabled and Elderly People*. London: OFTEL.

Tomaskovic-Devey, D. (1988) Poverty and social welfare in the United States. In: Tomaskovic-Devey, D. (ed) *Poverty and Social Welfare in the United States*. Boulder: Westview, pp. 1-26.

Tomkinson, M. (1993) Why it's last orders for licensees, *Sunday Express* 18 April, pp. 74-75.

Trent, R., Stout-Wiegand, N. and Smith, D. (1985) Attitudes toward new development in three Appalachian counties, *Growth and Change* 16 (4), pp. 70-86.

Tricker, M. and Martin, S.J. (1990) *An Evaluation of the Rural Development Commission's Support for Project Officer Posts, Final Report to the RDC*. Birmingham: Aston University.

Tucker, J. (1976) Changing patterns of migration between metropolitan and non-metropolitan areas in the United States: recent evidence, *Demography* 13, pp. 435-443.

Tykkylainen, M. and Neil, C. (1995) Socio-economic restructuring in resource communities: evolving a comparative approach, *Community Development Journal* 30 (1), pp. 31-47.

UK Round Table on Sustainable Development (1998) *Aspects of Sustainable Agriculture and Rural Policy*. London: UK Round Table on Sustainable Development.

United Nations (1992) *UN Convention on the Conservation of Biological Diversity*. New York: UN.

US Department of Commerce, Bureau of the Census (1991) *Current Population Reports, Series P-26*. Washington DC: US Government Printing Office.

US Department of Commerce, Economic and Statistics Administration (1992a) *1990 Summary Tape File 3, Population Tables*. Machine Readable Data Files. Minnesota. P6 Urban and Rural. Minneapolis: Machine Readable Data Center.

US Department of Commerce, Economic and Statistics Administration (1992b) Bureau of Census: *1990 Census of Population and Housing: Summary of Social, Economic, and Housing Characteristics*. Table 6, Employment Status and Journey to Work and Table 9, Income and Poverty Status in 1989.

US Department of Commerce, Economic and Statistics Administration (1994) Bureau of Census: *1992 Census of Agriculture: State Data, Minnesota*. Table 1, County Summary Highlights. Washington, DC: US Government Printing Office.

University of Maine Co-operative Extension (1994) *Getting It Done...LOCALLY*. Ideas to Action Handbook. Machias, Maine: University of Maine Co-operative Extension.

Urry, J. (1995) *Consuming Places*. London: Routledge.

Vogel, D. (1986) *National Styles of Regulation: Environmental Policy in Great Britain and the United States*. London: Cornell University Press.

Voth, D. (1989) Evaluation for community development. In: *Community Development in Perspective*. Ames, Iowa: Iowa State University Press.

Wachtel, H.M. (1971) Looking at poverty from a radical perspective, *The Review of Radical Political Economy* 3, pp. 1-19.

Wade, J.L. and Pulver, G.C. (1991) The role of community in rural economic development. In: Pigg, K.B. (ed) *The Future of Rural America: Anticipating Policies for Constructive Changes*. Boulder: Westview.

Walker, G. (1987) *An Invaded Countryside: Structures of Life on the Toronto Fringe*. Toronto: York University, Atkinson College, Geographical Monographs No. 17.

Walker, G.R. and Sheppard, P.J. (1997) Telepresence - the future of telephony, *BT Technology Journal* 15, pp. 11-17.

Walls, D. (1978) Internal colony or internal periphery? A critique of current models and an alternative formulation. In: Lewis, H., Johnson, L. and Askins, D. (eds) *Colonialism in Modern America: The Appalachian Case*. Boone: Appalachian Consortium Press, pp. 319-349.

Walzer, N. and Ching, P. (1994) Small towns in Illinois: assessing their economic development strategies, *Small Town* 25 (2), pp. 22-28.

Ward, N. (1993) The agricultural treadmill and the rural environment in the post-productivist era, *Sociologia Ruralis* 27, pp. 21-37.

Ward, N. and Lowe, P. (1994) Shifting values in agriculture: the farm family and pollution regulation, *Journal of Rural Studies* 10, pp. 173-184.

Ward, N., Lowe, P., Seymour, S. and Clark, J. (1995) Rural restructuring and the regulation of farm pollution, *Environment and Planning A* 27, pp. 193-211.

Ward, N., Clark, J., Lowe, P. and Seymour, S. (1998) Keeping matter in its place: pollution regulation and the reconfiguring of farmers and farming, *Environment and Planning A* 30, pp. 1165-1178.

Warren, R.L. (1977) *Social Change and Human Purpose.* Chicago: Rand McNally.

Water Authorities' Association and MAFF (1986) *Water Pollution from Farm Waste 1985 (England and Wales).* London: WAA.

Water Authorities' Association and MAFF (1987) *Water Pollution from Farm Waste 1986 (England and Wales).* London: WAA.

Water Authorities' Association and MAFF (1988) *Water Pollution from Farm Waste 1987 (England and Wales).* London: WAA.

Water Authorities' Association and MAFF (1989) *Water Pollution from Farm Waste 1988 (England and Wales).* London: WAA.

Watkins, C. (1984) The use of Forestry Commission censuses for the study of woodland change, *Journal of Historical Geography* 10 (4), pp. 396-406.

Watson, N., Mitchell, B. and Mulamottil, G. (1996) Integrated resource management: institutional arrangements regarding nitrate pollution in England, *Journal of Environmental Management and Planning*, 39(1), pp. 45-64.

Watts, H.D. (1987) Market areas and spatial rationalization: the British brewing industry after 1945, *Tidjschrift voor Economische en Sociale Geographie* 68, pp. 224-240.

Watts, H.D. (1991) Understanding plant closure: the UK brewing industry, *Geography* 76, pp. 315-330.

Weale, A. (1992) *The New Politics of Pollution.* Manchester: Manchester and New York University Press.

Weller, J. and Willetts, S. (1977) *Farm Wastes Management.* London: Crosby Lockwood Staples.

Wenger, G.C. (1990) *Keeping in Touch: Access to Cars and Telephones in Old Age*. Bangor: University College of North Wales Bangor Centre for Social Policy Research and Development.

West Virginia Bureau of Employment Programs (1993) *West Virginia County Profiles*. Charleston: Office of Labor and Economic Research.

West Virginia Chamber of Commerce (1987) *West Virginia Economic-Statistical Profile-1987*. Charleston: West Virginia Chamber of Commerce.

Whatmore, S.J. (1995) From farming to agribusiness: the global agro-food system. In: Johnston, R.J., Taylor, P.J. and Watts, M.J. (eds) *Geographies of Global Change: Remapping the World in the Late Twentieth Century*. London: Blackwell, pp. 3-49.

Whatmore, S.J., Munton, R., Marsden, T. and Little, J. (1987) Towards a typology of farm business in contemporary British agriculture, *Sociologia Ruralis* 27, pp. 103-122.

White, R. (1980) Poor men on poor land: the back-to-the-land-movement of the early twentieth century - a case study, *Pacific Historical Review 66* (3), pp. 105-131.

White, S. (1989) America's Soweto: population redistribution in Appalachian Kentucky, *Appalachian Journal* 16 (4), pp. 350-360.

Whitebloom, S. (1991) Brewers call time on tied houses, *Observer* 11 August, p. 26.

Whitley, E. (1992) Time, please, for the British Pub, *The Spectator* 268, 14 February, pp. 16-17.

Wilson, C.R. (1989) Growing tobacco is art for farmer, *Bristol Herald Courier* September 28, pp. 3B, 8B.

Winnifrith, J. (1962) *The Ministry of Agriculture, Fisheries and Food*. London: George Allen and Unwin.

Winter, M. (1996) *Rural Politics: Policies for Agriculture, Forestry and the Environment*. London: Routledge.

Wright, S. (1992) Rural community development: what sort of social change? *Journal of Rural Studies* 8 (1), pp. 15-28.

Yarrow, M. (1979) The Labor process in coal mining: struggle for control. In: Zimbalist, A. (ed) *Case Studies on the Labor Process*. New York: Monthly Review Press, pp. 170-192.

Zekeri, A. (1994) Adoption of economic development strategies in small towns, *Journal of Rural Studies* 10 (2), pp. 185-196.

Zimolzak, C. (1977) Changing ownership patterns in the West Virginia coal industry: oligopoly and its geographic impact. In: Adkins, H., Ewing, S. and Zimolzak, C. (eds) *West Virginia and Appalachia.* Dubuque: Kendall-Hunt, pp. 158-180.

Index

259